Quality Management in Scientific Research

Antonella Lanati

Quality Management in Scientific Research

Challenging Irreproducibility of Scientific Results

 Springer

Antonella Lanati
Valore Qualità
Pavia
Italy

ISBN 978-3-030-09573-4   ISBN 978-3-319-76750-5   (eBook)
https://doi.org/10.1007/978-3-319-76750-5

Printed on acid-free paper

This Springer imprint is published by the registered company Springer International Publishing AG part of Springer Nature.
The registered company address is: Gewerbestrasse 11, 6330 Cham, Switzerland

To Fabio and Chiara

Foreword

Antonella Lanati is a prominent member of the Research Quality Association (RQA); she is a prolific author of technical papers and books, mostly on the subject of quality in life sciences, biotechnology and pharmaceuticals.

The RQA is a global professional membership association supporting individuals concerned with the quality of research relating to pharmaceuticals, agrochemicals, and medical devices. The RQA is a leading provider of training, eLearning, publications, seminars, webcasts, conferences, news, and knowledge.

Antonella has worked across a broad spectrum of industries; she understands quality and how it is uniquely applied in each sector. She uses this experience and knowledge to encourage the reader to see beyond compliance and to explore a Total Quality approach.

Quality Management in Scientific Research—Challenging Irreproducibility of Scientific Results introduces Total Quality into the scientific research arena and, by using real examples, it explains how to utilise tried-and-tested tools and techniques in a research environment. The book is comprehensive and extremely well written; it enlightens the reader by introducing the Total Quality tools and techniques in a logical order.

The book takes the reader on a journey from the basics of quality, via ISO 9000 and the main reference standards for quality in non-regulated research, through methodologies such as TQM, Problem Solving and Project Management, onto FMEA, and finally into LEAN and 6-Sigma. At each stage, the author assumes that the reader knows nothing about the topic and goes about explaining how the methodologies can be utilised in scientific research.

Research is often an investigation into the unknown, sometimes addressing questions that have not been asked previously. The importance of quality management is acknowledged for routine analysis; it is less well established for research. *Quality Management in Scientific Research—Challenging Irreproducibility of Scientific Results* successfully blends research and quality management into an approach that will benefit scientific researchers and their organisations.

Research Quality Association

Contents

Abbreviations

AIAG	Automotive Industry Action Group (USA)
ALLEA	All European Academies
API	Active Pharmaceutical Ingredient
ATMP	Advanced Therapies Medicinal Product
CAPA	Corrective Action/Preventive Action
CB	Certification Body
CMC	Chemistry, Manufacturing, and Control
CR (curve)	Concentration-Response (curve)
CTQ	Critical to Quality
CWQC	Company Wide Quality Control
DIKW	Data–Information–Knowledge–Wisdom
DMAIC	Define–Measure–Analyze–Improve–Control
DoE	Design of Experiments
DPMO	Defects Per Million Opportunities
EFQM	European Foundation for Quality Management
ELN	Electronic Laboratory Notebook
EPA	Environmental Protection Agency (USA)
ESF	European Science Foundation
FDA	Food and Drug Administration (USA)
FMEA	Failure Mode and Effect Analysis
GCP	Good Clinical Practice
GLP	Good Laboratory Practice
GMP	Good Manufacturing Practice
GRP	Good Research Practice
ICH	International Conference on Harmonisation
ID	Interrelationship Diagram
IEC	International Electrotechnical Commission
ISO	International Organisation for Standardisation
IT	Information Technology
JCoPR	Joint Code of Practice in Research (UK)
KM	Knowledge Management
LIMS	Laboratory Information Management System
LSL	Lower Spec. Limit
NGS	Next Generation Sequencing

NIH	National Institutes of Health (USA)
OECD	Organisation for Economic Cooperation and Development
OFAT	One Factor at A Time
PDCA	Plan–Do–Check–Act (Deming Cycle)
PDSA	Plan–Do–Study–Act
PERT	Program Evaluation and Review Technique
PM	Project Management
PMBOK	Project Management Body of Knowledge
QA	Quality Assurance
QC	Quality Control
QMS	Quality Management System
QP	Quality of the Process
QPBR	WHO Handbook on Quality Practice in (Basic) Biomedical Research
QRS	Quality Research Service, Servei de Qualitat de la Recerca (University of Barcelona)
QS	Quality of the Scientific project
RCUCK	Research Council of the United Kingdom
RPN	Risk Priority Number
RQA	Research Quality Association (UK)
RRI	Responsible Research and Innovation
SIPOC	Supplier–Input–Process–Output–Customer
SOP	Standard Operating Procedure
SPC	Statistical Process Control
SQA	Society of Quality Assurance (USA)
TDR	Special Programme for Research and Training in Tropical Diseases (WHO)
TOC	Theory of Constraints
TPS	Toyota Production System
TQM	Total Quality Management
UAB	Universitat Autònoma de Barcelona
UB	University of Barcelona
UCL	University College London
UDE	Un-Desirable Effect
UKRIO	United Kingdom Research Integrity Office
USL	Upper Spec. Limit
VOC	Voice Of the Customer
VSM	Value Stream Mapping
WBS	Work Breakdown Structure
WHO	World Health Organisation
WP	Work Package

Introduction

In recent years, the scientific world has been experiencing a revolution: the attention of the scientific and social community is not focused solely on the results but also on the related issues of reliability, safety and efficacy of the discoveries as well as the efficient and effective use of resources. A strong debate is now going on about the reproducibility of scientific results in light of the increasing number of retractions of published articles, some of them because of obvious fraud, and others simply due to the lack of controls or poor practice. Starting from a couple of surveys made by Big Pharma about the reliability of scientific research products, it was soon pointed out in scientific journals [1–4] and even in the press [5–8] that, too often, trying to reproduce results published in peer-reviewed journals fails, even when obtained from large, experienced laboratories. While some cases of fraud have been discovered and have led to the recall of the papers, in other cases the irreproducibility could be attributed to either improper data management and processing or—more broadly—lack of good experimentation management. A single example is reported in *The Economist* [5]: a 2010 study published in *Science*, the prestigious American journal, was retracted a year later because of the strong criticisms from other geneticists: they complained about the different statistical treatments applied to the samples taken from centenarians and those from the younger control group. The authors justified the withdrawal admitting *technical errors* and *an inadequate quality-control protocol*. The number of retractions is growing faster than ever, tenfold over the past decade; this, however, represents no more than 0.2% of the 1,400,000 papers published annually in scientific journals [5]. These figures cannot account for the lack of reproducibility reported by Big Pharma, leaving the impression that the phenomenon deserves more attention and further investigation. Peer review has long been claimed to be the best way to judge the worth of scientific papers and to guarantee the validity of the scientific results presented for publishing. Experiments have been made by journalists and even by editors to test the real effectiveness of peer reviewing (see Box 2.1 for some examples). Overall, it is suggested that only a minority of reviewers carry out an in-depth analysis of the scientific results to identify possible errors. In some cases, they even appear to miss due

© Springer International Publishing AG, part of Springer Nature 2018
A. Lanati, *Quality Management in Scientific Research*,
https://doi.org/10.1007/978-3-319-76750-5_1

checks, either scientific or methodological. In any case, the peer review cannot be taken as the final, effective control for inaccuracies of the research project, although much has been done and is being done by the most important scientific journals to ensure the quality of what is published. Peer review has to be accompanied, indeed preceded, by a rigorous method for conducting the studies.

Funding agencies are interested in investing in research programmes that guarantee consistent results. The competition for research resources is hard and the funds obtained are often lower than the anticipated need. The selection is fierce and is made on the credibility of the applicants' previous scientific production. This aspect, paradoxically, constitutes also a risk for solid scientific products since the need to obtain resources may drive scientists to succumb to the temptation to over-interpret the results in order to ensure future funding for their own group [7, 9, 10].

In summary, scientists must consistently prove that they generate robust, reliable, traceable and repeatable data, in order to prove to funding agencies, and to the scientific and industrial world, that they deserve to be funded for their research. Once obtained, the—often limited—funds have to be strictly managed to make the most of them.

In order to guarantee reliable results, and improved consistency of data, while operating in a context of reduced funding, scientists need to acquire a new culture of management, more tools and specific training, so as not to be left floundering in a field that, apparently, has little to do with scientific research.

In this book, we intend to provide the basic knowledge to researchers who wish to improve their approach to non-regulated research management. The book explores different approaches to quality and proposes methodologies, principles and tools directly applicable to the field of non-regulated basic biological/biomedical research. Many examples of applications are shown, to allow the reader to become familiar with these subjects, and transfer them into their own field of experience.

Chapter 1 is devoted to the problem of reproducibility of scientific results and to how a Quality approach can give a substantial contribution to research. Chapter 2 introduces the reader to principles and basic concepts of Quality, paving the way for further discussion. Chapter 3 illustrates how Quality can be applied in basic biomedical research: from main international references to various aspects of research organisation and management. Chapters 4 and 5, respectively, are devoted to Quality tools and Quality methodologies, both enriched by examples and application in scientific research.

1.1 Acknowledgements

This book collects years of study and personal research, as well as experiences with different groups of scientists, all dedicated to delve into and apply quality disciplines in the management of biomedical studies. I have collaborated closely with valuable and Quality-devoted researchers whose contribution permeates the entire text. Here I would like to acknowledge all their contributions.

The experience with the Laboratory of Bioimaging of the San Raffaele Institute (OSR), Alembic, lasted 5 years. The development of the *ionChannelΩ* platform for drug screening was a challenging endeavour providing the opportunity to range from project management and software validation to image elaboration and application of quality methodologies such as FMEA and DoE. Fabio Grohovaz (Alembic-OSR and San Raffaele University) has been the visionary leader of the working group composed by Andrea Menegon and Nausicaa Mazzocchi (Alembic-OSR), Massimo Imberti e Alessandro Isernia (OpenSistemi), myself and Cecilia Poli (Assing/Engineers Professional Association for the Province of Pavia). The entire work has been rewarded with publication in Nature-Scientific Report [11].

The experience with the qPMO Project, later named Network, allowed all of us to demonstrate with tangible results how much the quality disciplines can be integrated in the scientific method, making the shift from bench to management. The qPMO Network was born in February 2012 and was funded by the CNR Life Sciences Department and the programme *FaReBio di Qualità*. It is composed by nine CNR researchers and technologists belonging to five different institutes, with my support as a quality consultant and with the coordination of Annamaria Kisslinger (IEOS, Institute for Endocrinology and Experimental Oncology). The qPMO Network is divided into four teams, taking in charge four Working Packages (WP1–4) on different aspects of TQM application to research: Giovanna L. Liguori and Giuseppina Lacerra (IGB, Institute of Genetics and Biophysics) and F. Anna Digilio (IBBR, Institute of Bioscience and BioResources) worked on Knowledge Management (WP1). Annamaria Kisslinger, Anna Mascia and Anna Maria Cirafici (IEOS) took care of management of experimental procedures (WP2). Antonella Bongiovanni and Marta Di Carlo (IBIM, Institute of Biomedicine and Molecular Immunology) designed the QMS for a research laboratory (WP3). Gianni Colotti (IBPM, Institute of Molecular Biology and Pathology) with, once again, G. L. Liguori (IGB) worked on management of multivariable assays (WP4). The qPMO Network received the contribution of other CNR researchers and technicians: Adriano Barra, Valeria Zazzu, Sara Mancinelli and Romeo Prezioso (IGB), and Loredana Riccobono, Maria De Bernardi, Letizia Anello, Luca Caruana and Alessandro Pensato (IBIM). Andrea Turcato, an engineer with a strong inclination for biomedical sciences, gave an invaluable contribution to the DoE studies.

Seeing the MoBiLab (Molecular Biodiversity Laboratory, CNR-IBIOM, Institute of Biomembranes, Bioenergetics and Molecular Biotechnologies) born and taking shape as a laboratory organised according to a management system has been a true professional pleasure. The MoBiLab is part of the European project *Lifewatch, E-science and technology infrastructure for biodiversity and ecosystem research* [1]. Researchers who work at the MoBiLab show dedication and great competence. Francesca De Leo leads the working group formed by Marinella Marzano and Caterina Manzari, with Giuseppe Sgaramella and Claudia Lionetti who also were part of the team. Their conviction in implementing a GLP-based management model as the primary tool to ensure excellence went far beyond the quality system: they

---

[1] LifeWatch: http://lifewatch.eu/What_is_LifeWatch. Accessed 13 Sep 2017.

also planned to realise a Laboratory Information Management System (LIMS) based on the MoBiLab GLP-like management system, aimed to become a solid reference in the NGS field. All this has been possible thanks to the strong motivation and support of Graziano Pesole, Head of the Institute.

The *Biotechnology Drugs and Advanced Therapies* team of the Italian pharmaceutical association *AFI Scientifica* gave me the opportunity to collaborate with esteemed experts from academia and industry, and share knowledge and methodological approaches on a variety of matters. We are still working with Marta Galgano as the team coordinator, Franco Pattarino (University of Piemonte Orientale), Marco Adami (AFI), Maria Luisa Nolli (NCNBio), Raffaella Coppolecchia and Achille Arini (Cerbios, Switzerland), Bice Conti and Cristina Bonferoni (University of Pavia), and Paola Scolari (Italfarmaco) on quality methodologies applied to the field of biological drugs and advanced therapies, in order to generate valuable references for this frontier science.

Last but certainly not least, I would like to thank experts and scientists who accepted to share with me their knowledge and contributed with short commentaries to this work: Rebecca Davies (University of Minnesota, USA), Shannon Eaker (GE Healthcare, USA), Melissa Eitzen (SQA, USA), Lucia Monaco (Telethon, Italy), Carmen Navarro Aragay (University of Barcelona, Spain), Franco Pattarino (University of Piemonte Orientale, Italy) and Daniele Zacchetti (San Raffaele Scientific Institute, Italy). I appreciate very much their competence, their immediate commitment to the project and their patient availability to tune their texts to the style and requirements of the project.

This work would have not been possible without the discrete, competent, patient, tireless help of Gabrielle Nasca Quadraccia, who contributed not only in transforming my language to a more fluent English, but also improving the readability and comprehensibility of the text, thanks to her degrees in biological sciences and her lively scientific curiosity.

1.2   Permissions

All WHO material is reprinted from WHO Handbook on Quality Practice in Basic Biomedical Research, available at http://www.who.int/tdr/publications/documents/quality_practices.pdf?ua=1, last accessed August 23rd 2017 (Copyright 2006).

The reproduction of ISO document excerpts from the book "Quality management principles Ed. 2015" and from the article "Manders & de Vries, Does ISO 9001 pay?—Analysis of 42 studies" has been authorised by UNI Italian National Unification Body on behalf of ISO International Organization for Standardization. The original versions are available in full on the ISO site https://www.iso.org/home.html.

References

1. Collins FS, Tabak LA. Policy: NIH plans to enhance reproducibility. Nature. 2014;505:612–3. https://doi.org/10.1038/505612a.
2. Freedman LP, et al. The economics of reproducibility in preclinical research. PLoS Biol. 2015;13(6):e1002165. https://doi.org/10.1371/journal.pbio.1002165.
3. Jasny BR, et al. Fostering reproducibility in industry—academia research—sharing can pose challenges for collaborations. Science. 2017;357(6353):759–61. https://doi.org/10.1126/science.aan4906.
4. Ioannidis JPA. Why most published research findings are false. PLoS Med. 2015;2(8):e124.
5. Trouble at the lab—scientists like to think of science as self-correcting. To an alarming degree, it is not. The Economist. 2013. http://www.economist.com/news/briefing/21588057-scientists-think-science-self-correcting-alarming-degree-it-not-trouble. Accessed 13 Sep 2017.
6. How Science goes wrong—scientific research has changed the world. Now it needs to change itself. The Economist. 2013. http://www.economist.com/news/leaders/21588069scientificresearchhaschangedworldnowitneedschangeitselfhowsciencegoeswrong. Accessed 13 Sep 2017.
7. Achenbach J. The new scientific revolution: reproducibility at last. The Washington Post 2015. https://www.washingtonpost.com/national/health-science/the-new-scientific-revolution-reproducibility-at-last/2015/01/27/ed5f2076-9546-11e4-927a-4fa2638cd1b0_story.html?utm_term=.61cd223ff312. Accessed 26 Sep 2017.
8. Naik G. Scientists' elusive goal: reproducing study results. The Wall Street Journal 2011. https://www.wsj.com/articles/SB10001424052970203764804577059841672541590. Accessed 26 Sep 2017.
9. Eanes Z. NC State University group accused of falsifying research. dailytarheel.com. 2014. http://www.dailytarheel.com:8080/article/2014/02/nc-state-university-group-accused-of-falsifying-research. Accessed 13 Sep 2017.
10. Ségalat L. System crash. EMBO Rep. 2010;11(2):86–9. https://doi.org/10.1038/embor.2009.278.
11. Menegon A, et al. A new electro-optical approach for conductance measurement: an assay for the study of drugs acting on ligand-gated ion channels. Sci Rep. 2017. https://doi.org/10.1038/srep44843.

References

How Quality Can Improve Reproducibility 2

2.1 The Unknown of Reproducibility

The problem of lack of reproducibility of scientific results is showing in the last years—even months—as one of the most relevant emerging topics in scientific literature. A quick search in PubMed with the keyword *reproducibility* and *research* leads to the data illustrated in Fig. 2.1. The number of scientific papers containing these two keywords in the title has grown in the last 5 years from very few to more than 20 per year. This trend witnesses the awareness of the scientific world on the consequences of the publication of poor results: they range from the rising costs for industrialisation, which must include a rigorous replication before starting a new treatment development, to delays in the development of disease treatments. Adverse consequences include not only the influence on distribution of research funding and the failure to meet the community's expectations, but also a detriment to credibility of science itself. This latter aspect is also subtly harmful from a social point of view: as scientific breakthroughs increase confidence in science [1], by the same token any public evidence of misconduct or unfulfilled expectations undermines confidence in science, leading to populisms and laying ground for false beliefs. Vaccine controversies as well as *unconventional* cancer treatments based on forms of alternative medicine are some of the overt examples of such social aberrations.

2.1.1 Irreproducibility and Retractions

As recalled in the Introduction, the problem of reproducibility was raised by Big Pharma, such as Bayer and Amgen, after they invested in large teams to confirm results published in scientific papers regarding new strategies for cancer treatment. More specifically, they claimed that they were able to replicate less than 9% of the results [2–5]. Also of note, as reported in a Nature Reviews paper [6], is the success rate of Phase 2 trials which dropped from 28% in 2006–2007 to just 18% in 2008–2010.

© Springer International Publishing AG, part of Springer Nature 2018   7
A. Lanati, *Quality Management in Scientific Research*,
https://doi.org/10.1007/978-3-319-76750-5_2

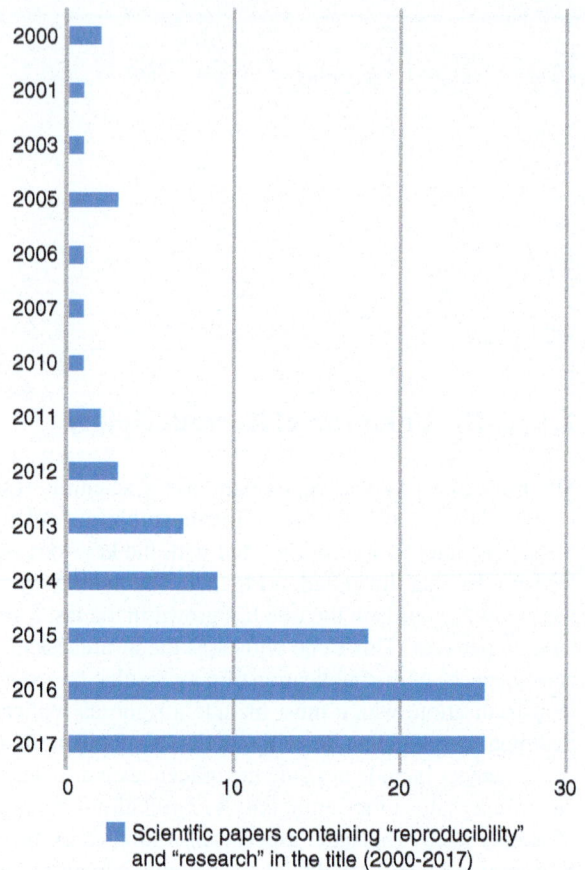

Fig. 2.1 Scientific papers containing *reproducibility* and *research* in the title (2000–2017). The search of the PubMed database for papers containing the words *reproducibility* and *research* in the years since 2000 till 2017 shows a more than tenfold increase, from 1 or 2 in 2000/2001 to 25 in 2016 and first two quarters of 2017

Scientific papers containing "reproducibility" and "research" in the title (2000-2017)

Similar initiatives were taken also by scientists in the attempt to reproduce results already published in their field of study. An example is provided by Héroux et al. [7] who published a survey in which they examined published papers and the results of interviews with colleagues. The authors found the field *not immune to issues of reproducibility*, sometimes even recognised by the authors themselves, mainly due to questionable research practices and publication biases that compromise the value of the research. Another example is given by Mobley et al. [8], who performed a similar survey via a questionnaire distributed to the faculty and trainees at MD Anderson Cancer Center, Houston, TX, USA. About 55% of survey respondents confirmed that they were not able to reproduce findings from a published paper; however, one-third of them succeeded in reproducing the results after being in contact with the authors.

Freedman et al. [9] summarised in a Pareto chart the estimation of data irreproducibility according to various authors (Fig. 2.2), showing that it can reach values of up to 89%. However, they also observed that the authors of the papers cannot be considered entirely liable for irreproducibility, since errors can be made also in the

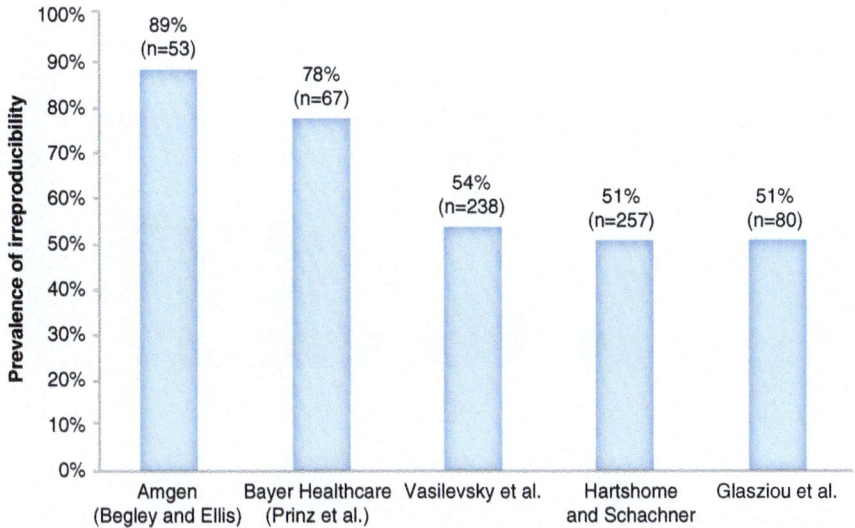

Fig. 2.2 Studies reporting the prevalence of irreproducibility. The chart illustrates the percentage of non-reproducibility and the sample size declared by several authors in the attempts to replicate published results (adapted from [9])

attempt to reproduce data. In any case, it emerges that the information deduced from an article is often insufficient to allow easy and full replication of the experiments.

In the absence of consistent data, we can try to extract some information from an indirect measure of irreproducibility, namely retractions of scientific papers,[1] assuming that it may be significantly affected by the authors' admission of errors. This information must be treated with great caution, for several reasons: first, retractions can also be due to the identification of misconduct, discovered with a delay of up to several years. This metric, however, is expected to grossly underestimate the phenomenon, virtually being restricted to the most important journals, which are under the magnifying lens of researchers for the value of the published material and are more driven to self-control and correction because of the awareness of the scientific value of what they publish. The problem of retractions has also gained space in the press, where it may receive excessive emphasis for the sake of sensationalism. In the light of these premises, we can analyse some data from press and scientific papers.

The New York Times [10] presented a statistical analysis of retractions of scientific papers from the year 2000 to 2009, performed by two researchers, F. C. Fang and A. Casadevall, on the PubMed database. The bar graph in Fig. 2.3 shows a tenfold increase. The authors also broadly subdivided the reasons for retractions into three classes: *fraud or fabrication*, *scientific mistake* and *other*. As already

---

[1]A useful source of information regarding paper retractions can be found in a watchdog blog, Retraction Watch (http://retractionwatch.com/), which provides up-to-date information and commentaries.

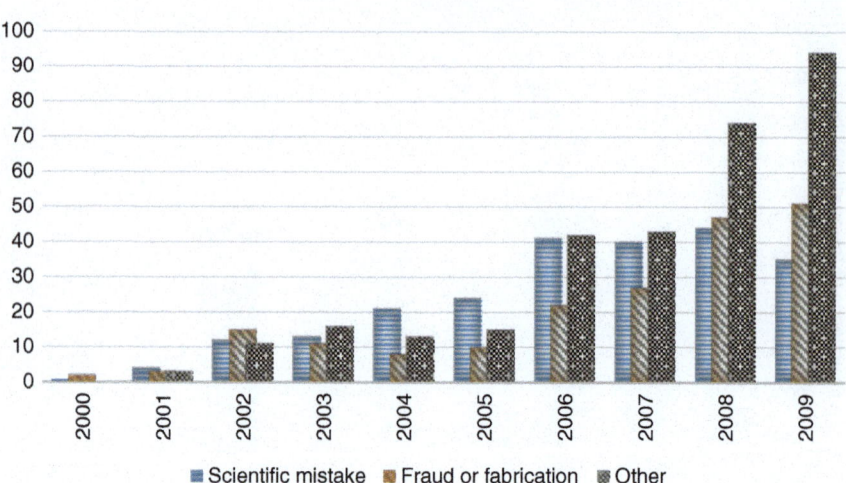

Fig. 2.3 Retractions of scientific papers (2000–2009). The study of paper retractions on the PubMed database found a very significant increase between years 2000 and 2009. Reasons for retraction are classified according to *Scientific mistake* (tot. 235), *Fraud or fabrication* (tot. 196) and *Other* (tot. 311)—data from [11] and [10]

noted, fraud is not the major issue for retraction; rather it is diminishing with time. While scientific errors remain a stable percentage, an alarming increase is shown for the undefined *other* class. It is very likely that study management problems are hidden under this definition, together with other, less defined, misconduct problems.

On the other hand, a survey performed by Steen [11] highlights an increasing withdrawal rate for misconduct, showing that more than 30% of retractions were due to scientific error—including inability to reproduce results—and represented the most common reason, while another 18.1% was attributed to unidentified causes and thus unclassified. Of note, the retractions due to reason other than fraud were published in the journals with the higher mean impact factor, suggesting a higher level of attention by the editors and readers of the most important publications.

2.1.2   The Economics of Poor Reproducibility

The problem of reproducibility, besides undermining the credibility of science and damaging subsequent research,[2] has further drawbacks, including an economic impact that cannot be neglected. The paper from Freedman et al. [9] presents a clear picture of the economics of irreproducibility. The authors, based on estimates of irreproducibility that claim figures ranging from 51 to 89% (see Fig. 2.2) and performing a statistical evaluation, came to the conclusion that a conservative

---

[2] Montgomery and Oliver [12] model the consequences on the international scientific community of a false science article, which contaminates the knowledge system when it is kept as valid, to the mechanisms of spreading of a virus.

estimated rate can be set at 50%. Applying this percentage to the number of US grants invested yearly on preclinical research—approximately $56B from the pharmaceutical industry, federal government, non-profits and academia—gives the impressive figure of $28B spent each year in studies whose results cannot be replicated, and this for the US alone. As pointed out by the same authors, these are not the only costs of lack of reproducibility: financial and opportunity costs are other sources of inefficacy, just to cite a few; however, one of the most significant is the cost for pharmaceutical industries to check the reliability of previous studies. As testified in a report issued by PhRMA[3] and cited by Freedman, each study replication may last from 3 to 24 months and may require investments ranging from US$500,000 to US$2,000,000.

Overall, an exhaustive assessment of the lack of research reproducibility, as well as of its economic consequences, is still difficult to outline. Nevertheless, the estimates are significant enough to raise the attention of the scientific world, and to stimulate the analysis of the possible causes and the search for suitable countermeasures.

2.1.3   Reasons and Solutions

There is no doubt that this increasing trend in irreproducibility has accompanied the escalation in the complexity of scientific studies involving the cooperation of many researchers, experts in different areas, the management of large data sets, the mastery of sophisticated statistical techniques and the use of high-tech tools.

In the numerous recent papers on the subject, many authors illustrate several reasons to justify the lack of reproducibility, including poor readability of papers [13]; lack of formal presentation of data [14]; incomplete communication and sharing of methods, data and materials between research and industry [15]; lack of transparency [16]; low statistical power [16, 17]; and perverse incentives (see, among others, [18]). Two leaders of US National Institutes of Health (NIH) [19] attribute the causes of irreproducibility to several aspects, such as a lack of training of young researchers in experimental design, the increasing trend to make grand statements instead of presenting technical details, and poor supply of data and information about experimental design, often driven to keep a competitive edge. Freedman et al. [9] define four categories of error that contribute to irreproducibility: (1) biological reagents and reference material (36.1% of the total); (2) study design (27.6%); (3) data analysis and reporting (25.5%); and (4) laboratory protocols (10.8%).

Three of these aspects deserve special attention: training of young researchers, perverse incentives and communication between industry and research. Training will be discussed below. Researchers have long been aware that the number and the importance of their publications have a great influence on their career and their ability to obtain funds. The term *Publish or Perish*, which was coined in the early 1900s,

---

[3] PhRMA (The Pharmaceutical Research and Manufacturers of America) is an association of leading US biopharmaceutical researchers and biotechnology companies.

synthesizes the pressure to publish in academia. Unfortunately, under pressure researchers are prone to publish inaccurate data, flawed results and forced conclusion. Often too little time is devoted to confirming results, executing appropriate statistical analysis and performing proper controls. The inclination of funding agencies to overvalue researches published in high-IF journals as well as the incentives that some research institutions grant to authors of such publications are also incentives that may accelerate misleading publishing. On the other hand, many journals also collude with their implicit requests for rapid publication, at the expense of proper peer review and editorial assessment [20]. This rush to high-impact publication has also restrained the dissemination of negative results, which would be useful for scientists in avoiding unfruitful experimentations, thereby saving time and money. Publication of negative results has decreased from the 30% in 1990 to 14% in 2007 [3]. The reasons for this decline are synthetised by F.C. Fang, stating that the focus is not about the quality of the research, but on the need *to put the best face on everything* [10]. Overall, the increasing perception of a progressive drift in the standards of science should raise awareness of the importance of a more stringent application of ethical, together with more robust, research management and methodological experimental conduct.

The difficulty to establish mutual understanding between people of different professions is widely recognised, yet the problem of communication between academic research and industry has received little attention when addressing the problem of reproducibility. The two realities are part of the same chain, but they have very different goals and approaches. Academia has strong incentives to generate and provide confirmation of new ideas, disseminate them and obtain grant money from them. Industry gathers new knowledge or data and works to develop them into producible and saleable products. Scientists from academia often produce and describe results of very narrow interest, which apply only to well-defined conditions. Scientists from industry look for the validation of results and for the most effective and efficient scale-up to a production framework. The critical step in the progression of this research and development chain is represented by the discontinuity between the two worlds. Besides aims, materials and protocols, approaches and time constraints, also language and methods—both scientific and managerial—are different. The regulated environment of the pharmaceutical industry is "accustomed" to Quality terms and concepts, and relies on methodologies, techniques and tools barely familiar to basic research. A proper translation and adoption of a regulated culture in the scientific field would not only give a substantial contribution to the management of science, but also ease the communication, familiarising the lexicon and disseminating some key concepts. Most importantly, this positive *contamination* should help making research more reliable by introducing quantitative criteria in the evaluation of research outputs.

In general, the concept that science is self-correcting should be reconsidered. Even the idea that good science is preserved by the well-known mechanism of peer review has been shown to suffer from considerable weaknesses and has been challenged several times. In Box 2.1 are reported two experiences, one from the editor of a renowned scientific journal and the other from a *fake* author. Editorial commitment to publish valuable research and results is already in place. Nature, one of the most respected scientific journals, has dedicated a special section of its website to *Challenges in Irreproducible Research* [21]. This special section collects contributions on the theme of reproducibility, with the aim of promoting awareness within the whole of the scientific community, including journals, scientists, institutions and funders. This initiative was followed in 2013 by an announcement [22] stating that more restrictive controls and editorial measures were to be put in place by the journal to contribute to the reduction of irreproducibility: among others, the abolition of the length of method section, the introduction of a checklist for authors to help editors and reviewers to verify experimental designs, the scrutiny by statistical experts and the request to authors to provide more raw data. The Committee on Publication Ethics (COPE) has issued a code addressed to editors to promote integrity in research publication that also advises editors on how to handle cases of research and publication misconduct. The original text of the Code of Conduct for Journal Editors [23] and that for Publishers [24] can be found on the website of the association. In brief, it stresses the responsibilities of editors and publishers towards quality, freedom and correctness of the published material, giving some suggestions regarding the relationships with readers, authors and reviewers, and recommending special care for the peer review process and complaint management.

> **Box 2.1: A Check on Peer Review Reliability**
> In a 2013 article in The Economist [2], a couple of alarming examples are reported. The first is about a biologist at Harvard, John Bohannon, who—under a pseudonym—sent a paper regarding the effects of a chemical derived from lichen on cancer cells to more than 300 journals. The work contained blunders in the study design, in the analysis and in the interpretation of results. Neither the fictional author nor the clumsy errors prevented more than 150 journals from accepting the paper. Back in the late 1990s, another well-known example involved the editor of the BMJ, Fiona Godlee, who sent to more than 200 of the journal's regular reviewers a fake paper containing eight mistakes in study design, analysis and interpretation. Some reviewers did not even find one, most found two on average and no one intercepted them all.

Finally, it should be noted that also governments and grant agencies are taking steps to counteract the problem of irreproducibility of data, with particular attention to paper retractions, even though the action of claiming back money granted on false or wrong assumptions is still very rare.

Table 2.1 Causes of irreproducibility of scientific results

Misconduct, fraud	Ethics
Pressure for publishing	
Reluctance to publish negative results	
Inadequate study design and management	Process control
Managing of experimentation: • Many uncontrolled variables • Materials non-characterised • Contaminated samples • Inaccurate data management • Non-standard /inaccurate/wrong laboratory protocols • Inadequate quality control	
Missing or inaccurate documentation	
Low statistical power/lack of statistical expertise	
Communication with downstream research	
Effectiveness of peer review	Product control

The causes can be grouped in three classes, regarding ethics, process control and product control

Reasons for irreproducibility are multifarious and most of them reflect good faith and are not in contrast with a conscientious approach to work, making the difference with fraud and misconduct. Regardless of the reasons, we should indeed consider if we can afford such practices. If we synthetise and search for common denominators, the most important reasons can be grouped into three classes, as shown in Table 2.1. Two classes refer to the control of the process—namely how well the process is performed—and the control of the product—namely what is produced and how well. When dealing with the control of the research process, we do not refer simply to the experimentation, but to the whole organisation: management of the study, organisation and responsibilities, internal and external communication, training and knowledge, protocols and methods, documentation, procurement and materials, equipment and other resources, etc. What we are outlining in this way is the scope of a quality management system, which has another characteristic of utmost importance for scientific research: it aims at continuous improvement.

The third class of causes of irreproducibility in Table 2.1 is about ethical aspects, namely fraud and misconduct, which are beyond the aim of this book. However, it is worth noting that a control of the process and of the research outcome would also greatly help the group leader to keep under control the integrity of the work performed by the research staff. Moreover, and as funding institutions and publishers are beginning to ask more frequently, having drafted and declared a clearly planned study, including criteria for data analysis, experimental conditions, and sought and expected results, leaves less room for fraudulent manipulation of data and results.

It is interesting to analyse what was done in different fields, and how successful it was in increasing the reliability and repeatability of results. If we have a look at computer science and drug development—to name two—we can see that both share a high level of standardisation, quality practices and auditing: W3C and IETF are two examples for the former, and good laboratory practice (GLP) for the latter.

The prevention of poor management in research must start with education. The insufficient training of students in experimental design methods is one of the reasons highlighted by Collins and Tabak of NIH [19]. In their interview to the New York Times [10], F.C. Fang and A. Casadevall state that to change the system, it is necessary to start from education, by giving graduate students a better understanding of science's ground rules. R. Davies goes beyond, in his approach to good research [25], recommending young scientists be trained in research data management and quality assurance. The international institution EMBO, based in Germany, has been offering for many years a broad-based training programme to laboratory managers, including matters strictly related to the good management of research, even if not explicitly linked to quality. US NIH has recently implemented an intervention in order to tackle the problem: *Because poor training is probably responsible for at least some of the challenges, the NIH is developing a training module on enhancing reproducibility and transparency of research findings, with an emphasis on good experimental design* [19]. The training will be mandatory for their own researchers and training documentation will be published on the NIH website. Despite these clear indications, most educational institutions still ignore the quality training offer and leave current and future researchers expert in science but lacking the skills for an efficient and effective management of experimental practice.

In order to effectively introduce quality in the scientific world, it is necessary that the main stakeholders act with coordinated actions:

• University/academia, by implementing the training of researchers
• Research institutions, by issuing internal rules
• Funding agencies, by requiring research to be conducted in a regulated manner
• Governments, by adopting reward policies
• Scientific journals, by applying rigour in peer review

A key role in the dissemination of the quality approach within the scientific community is played by associations, such as RQA based in the UK and SQA based in the USA, both having an international vocation and counting members from all over the world. The Research Quality Association (RQA), founded in the UK as a reference for national regulatory standards in pharma, has quickly become an international player for quality in research, disseminating its learning products throughout the world, and accepting members from over 50 different countries. RQA publishes guidelines and handbooks to provide clear references for the application of quality principle and ISO model to non-regulated research. Several references to RQA can be found throughout the book. Box 2.2 illustrates the initiatives of the SQA, with a special focus on activities for disseminating quality in scientific research in universities.

Box 2.2: The Society of Quality Assurance (SQA)

The Society of Quality Assurance (SQA) is a global professional membership organisation, consisting of over 2000 professionals from more than 40 countries, dedicated to promoting and advancing the principles and knowledge of quality assurance essential to human, animal and environmental health. SQA is committed to promoting global awareness and understanding of regulations, industry standards, and consensus and guidance documents supporting the development and approval of chemicals, pesticides, drugs, biologics and medical devices. SQA provides professional development, education and training and creates collaborative relationships with governmental authorities, professional organisations, academic institutions, and business and industry.

Within the organisation, members are able to join smaller specialty sections that focus on specific areas aligned with their interests and expertise. The University Specialty Section (USS) focuses on programmes established within academic settings that support the development of regulated products and promote data integrity and research reproducibility. USS members provide training for university faculty and students regarding quality assurance, regulated research and good documentation practices in order to promote quality in research and regulated studies. Members share their respective approaches via recurring scheduled teleconferences; collaborate to provide education and presentations at national and international meetings; support career development for graduate students and postdoctoral fellows; serve as mentors in the SQA mentoring programme; and participate in exchange programmes where members travel to other academic campuses to share knowledge and ideas.

Information describing education and training (including face-to-face and online programmes), SQA's Registered Quality Assurance Professional program, the STEM Committee (promoting the integration of quality in science to students), the mentoring program and other resources can be found at www.sqa.org.

Melissa Eitzen, MS, RQAP-GLP
2017 President
Society of Quality Assurance

2.2   What Is Quality

We are introducing quality as the standard approach to guarantee the reproducibility of an outcome. But what exactly is *quality*? If we want to simplify the concept, working in quality is *working well*, i.e. managing well your own work. This cannot be done without making explicit reference to good practices, a compendium coding experiences of decades and in some cases even centuries, and without verifying their application. We can find many examples throughout human history. In the ancient Chinese Empire under the Shang dynasty (14th–11th centuries BC), inspectors were sent to the suppliers of the Imperial House to verify that the goods destined for the Emperor were constructed according to precise rules; today we would call it an *audit*. In the Italian Middle Ages (thirteenth century), the Corporations of Arts and Crafts took care and protected the quality of manufactured goods. The internal rules imposed strict controls over the use of raw materials, working tools and processing techniques, also fighting fakes, i.e. products that did not meet the quality standards of the associations.

The quality scholars E. Deming and J. Juran, in the middle of the last century, distinguished between quality of production and quality of management. Indeed, there is a substantial difference between *making* and *managing*: *making* regards the matter, the *what*, and *managing* refers to the organisation, the *how*. Thus, there is the need for methods and tools helping to do the job better, and tools helping in organising and supporting its execution.

In basic scientific research, the quality of management has lagged behind, and much knowledge can be drawn from fields where this matter has been, for historical and economic reasons, greatly developed. The same quality of *doing*, a concept familiar and very dear to researchers, can take advantage of some of the tools developed in disciplines related to the industrial world.

2.3 Quality for Research

Quality management in scientific research is emerging as an essential tool to ensure valuable, robust and dependable outcomes, within a framework of best practice. Quality disciplines have been widely used for decades in industrial and business fields. The central concept in the quality approach is the importance of the result for the final users. This focus easily leads to effectiveness – intended as a guarantee of results, prevention and safety – and to efficiency – intended as rigorous resource management and minimization of waste. In the commercial world, these aspects are strong drivers of strategy development and related tools. Under the aegis of quality, significant methodologies and approaches have been developed, covering almost all areas of business and production: from management systems to production control and from project and innovation management to statistical tools. However, only recently quality management and approaches are receiving proper attention in scientific research, especially in the life sciences, because of the prejudice that they are an impediment to creativity.

Scientific research culture and method pay great attention to the quality of the data and even more to their complex, multi-parametric analysis: this aspect is increasing in importance as the quantity of data to be analysed is rapidly growing. Less attention is generally devoted to how research is managed. Quality management is based on two main principles: process—how I get there—and product—what I achieve—and, consequently, attention is focused to efficiency and effectiveness. While the evaluation of research products has been widely adopted, monitoring of the research process is less frequent, even though it is essential in ensuring sound and reliable results. In most basic research environments a resistance to the introduction of quality assurance programs is not uncommon, and the appreciation of their potential benefits is not widely shared. The resistance mainly stems from fear of the complexity and of the extra time necessary to adopt new formalised approaches. However, the researchers who decide to apply quality principles and tools soon become convinced of the overall advantages. A paper recently published in Nature [26] reports some examples of the introduction of quality assurance practices, overcoming scepticism and reluctance, to achieve remarkable outcomes.

A proper deployment of the quality approach has long found its application in drug discovery and development through the standards of good practice (good laboratory practice, GLP [27]); in contrast, basic biological research is still lacking a widespread culture of quality. Moreover, it should be noted that even GLP, which controls and rules preclinical scientific experimentation, leaves unaddressed most management issues as well as strategies for continuous improvement. This was the case in good manufacturing practice (GMP), the international standards for pharmaceutical production, now facing an evolution to a quality management system driven by the International Conference for Harmonization [28]. The EU Commission has recently issued a document regarding the *Responsible Research and Innovation—RRI* [29]. The scope of such document has greater breadth than the sole quality approach: it bears witness to the EU Commission's interest, strategy and consciousness of the social impact of scientific research. The RRI suggests ways to strengthen a responsible approach to science, with a list of gradual interventions. In this light, recent EU calls for research projects either require or strongly suggest a transversal quality management work package, to ensure the training, control (monitoring and evaluation) and application of quality tools and methodologies. Recent examples of the application of International Organization for Standardization—ISO 9001 standard [30] in research structures have indicated many advantages in terms of governance, control, efficiency and results [31, 32]; similar benefits can be obtained by the application of quality principles in peer review processes [33].

Significant quality approaches to scientific research, even if not referring to ISO standards, are becoming more frequent in Europe and throughout the world. Different quality models have been proposed and in some cases a quality approach has become a requirement for all project proposals submitted for funding: in the UK, the Joint Code of Practice (for quality) in Research—JCoPR is strongly suggested as the standard by which to carry out scientific research for the main funding agencies. *The JCoPR sets out standards for the quality of science and the quality of research processes that contractors who carry out research on behalf of DEFRA,*[4] *and other signatory organisations, must follow. This helps ensure the aims and approaches of research are robust, and gives confidence that processes and procedures used to gather and interpret the results of research are appropriate, rigorous, repeatable and auditable* [34]. In 2008 RQA published a brief guideline [35] and in 2013 a handbook [36] to provide clear references for the application of quality

---

[4] Department for Environment, Food and Rural Affairs—UK.

principle and ISO 9001 model to non-regulated research. The most comprehensive reference for quality in research is still the Quality Practices in Basic Biomedical Research (QPBR) Handbook [37] issued in 2010 by WHO-TDR.[5] In the last few years, some universities and councils in the UK have produced their own good research practice, followed by other similar institutions in Europe, such as the University of Barcelona. The content of these guidelines will be analysed in depth in Chap. 4 of this book.

2.4   Research Integrity

At the end of this introduction, the so-called *research integrity* deserves special consideration, due to its great international importance and its deployment in national and supranational institutions. Research integrity can be defined as the trustworthiness of research due to the soundness of its methods and the honesty and accuracy of its presentation. This definition is drawn from the Singapore Statement on Research Integrity [38], released on 22 September 2010, which has been the product of the collective effort of a small committee of researchers, funders and research institutions, and was finalised thanks to the participants in the 2nd World Conference on Research Integrity. The principles of the Statement are *honesty* in all aspects of research, *accountability* in the conduct of research, *professional courtesy and fairness* in working with others and *good stewardship* of research on behalf of others. These principles are outlined in 14 responsibilities, regarding integrity, adherence to regulations, research methods, research records, research findings, authorship, publication acknowledgement, peer review, conflict of interest, public communication, reporting irresponsible research practices, responding to irresponsible research practices, research environments and societal considerations. Most of these principles will be addressed in this book.

References

1. Hillgard J, Jamieson KH. Does a scientific breakthrough increase confidence in science? News of a Zika vaccine and trust in science. Sci Commun. 2017;39(4):548–60. https://doi.org/10.1177/1075547017719075.
2. Trouble at the lab—scientists like to think of science as self-correcting. To an alarming degree, it is not. The Economist. 2013. http://www.economist.com/news/briefing/21588057-scientists-think-science-self-correcting-alarming-degree-it-not-trouble. Accessed 13 Sep 2017.
3. How Science goes wrong—scientific research has changed the world. Now it needs to change itself. The Economist. 2013. http://www.economist.com/news/leaders/21588069scientificrese archhaschangedworldnowitneedschangeitselfhowsciencegoeswrong. Accessed 13 Sep 2017.
4. Achenbach J. The new scientific revolution: reproducibility at last. The Washington Post. 2015. https://www.washingtonpost.com/national/health-science/the-new-scientific-revolution-repro-ducibility-at-last/2015/01/27/ed5f2076-9546-11e4-927a-4fa2638cd1b0_story.html?utm_term=.61cd223ff312. Accessed 26 Sep 2017.

---

[5] World Health Organisation (WHO) and its Special Programme for Research and Training in Tropical Diseases (TDR).

5. Naik G. Scientists' elusive goal: reproducing study results. The Wall Street Journal. 2011. https://www.wsj.com/articles/SB10001424052970203764804577059841672541590. Accessed 26 Sep 2017.

6. Arrowsmith J. Trial watch: phase II failures: 2008–2010. Nat Rev Drug Discov. 2011;10:328–9. https://doi.org/10.1038/nrd3439.

7. Héroux ME, et al. Questionable science and reproducibility in electrical brain stimulation research. PLoS One. 2017;12(4):e0175635. https://doi.org/10.1371/journal.pone.0175635.

8. Mobley A, et al. A survey on data reproducibility in cancer research provides insights into our limited ability to translate findings from the laboratory to the clinic. PLoS One. 2013;8(5):e63221. https://doi.org/10.1371/journal.pone.0063221

9. Freedman LP, et al. The economics of reproducibility in preclinical research. PLoS Biol. 2015;13(6):e1002165. https://doi.org/10.1371/journal.pbio.1002165.

10. Zimmer C. A sharp rise in retractions prompts calls for reform The New York Times. 2012. http://www.nytimes.com/2012/04/17/science/rise-in-scientific-journal-retractions-prompts-calls-for-reform.html. Accessed 2 Oct 2017.

11. Steen RG. Retractions in the scientific literature: is the incidence of research fraud increasing? J Med Ethics. 2011;37:249–53. https://doi.org/10.1136/jme.2010.040923.

12. Montgomery and Oliver. Conceptualizing fraudulent studies as viruses: new models for handling retractions. MINERVA. 2017;55(1). https://doi.org/10.1007/s11024-016-9311-z

13. Plavén-Sigray P, et al. The readability of scientific texts is decreasing over time. eLife. 2017;6:e27725. https://doi.org/10.7554/eLife.27725.

14. Pasquier T, et al. Comment: if these data could talk. Scientific Data. 2017;4:170114. https://doi.org/10.1038/sdata.2017.114.

15. Jasny BR, et al. Fostering reproducibility in industry—academia research. Science. 2017;357(6353):759–61.

16. Gilmore RO, et al. Progress toward openness, transparency, and reproducibility in cognitive neuroscience. Ann N Y Acad Sci. 2017;1396(1):5–18. https://doi.org/10.1111/nyas.13325.

17. Dumas-Mallet E, et al. Low statistical power in biomedical science: a review of three human research domains. R Soc Open Sci. 2017;4:160254. https://doi.org/10.1098/rsos.160

18. Furst T, Strojil J. A patient called Medical Research. Biomed Pap Med Fac Univ Palacky Olomouc Czech Repub. 2017;161(1):54–7.

19. Collins FS, Tabak LA. Policy: NIH plans to enhance reproducibility. Nature. 2014;505:612–3. https://doi.org/10.1038/505612a.

20. Bauchner H. The rush to publication—an editorial and scientific mistake. JAMA. 2017;318(12):1109–10.

21. Nature: challenges in irreproducible research. http://www.nature.com/news/reproducibility-1.17552. Accessed 13 Sep 2017.

22. Editorial. Announcement: reducing our irreproducibility. Nature. 2013;496:398. https://doi.org/10.1038/496398a.

23. COPE: code of conduct and best practice guidelines for journal editors. 2011. https://publicationethics.org/files/Code%20of%20Conduct_2.pdf. Accessed 13 Sep 2017.

24. COPE: code of conduct for journal publishers. 2011. https://publicationethics.org/files/Code_of_conduct_for_publishers_Mar11.pdf. Accessed 13 Sep 2017.

25. Davies R. Good research practice: it is time to do what others think we do. Quasar-RQA. 2013;124:21–3.

26. Baker M. How quality control could save your science. Nature. 2016.; http://www.nature.com/news/howqualitycontrolcouldsaveyourscience1.19223

27. OECD: OECD series on principles of good laboratory practice (GLP) and compliance monitoring (1995–2006). http://www.oecd.org/chemicalsafety/testing/oecdseriesonprinciplesofgoodlaboratorypracticeglpandcompliancemonitoring.htm. Accessed 13 Sep 2017.

28. US Food and Drug Administration: ICH international conference on harmonization of technical requirements for registration of pharmaceuticals for human use Q10 pharmaceutical quality system. 2009. https://www.fda.gov/downloads/drugs/guidances/ucm073517.pdf. Accessed 13 Sep 2017.
29. Directorate-General for Research and Innovation Science in Society: options for strengthening responsible research and innovation. Report of the expert group on the state of art in Europe on responsible research and innovation. 2013. https://ec.europa.eu/research/science-society/document_library/pdf_06/options-for-strengthening_en.pdf. Accessed 13 Sep 2017.
30. ISO 9001:2015 Quality management systems—requirements.
31. Poli M, Pardini S, Passarelli I, Citti I, Cornolti D, Picano E. The 4A's improvement approach: a case study based on UNI EN ISO 9001:2008. Total Qual Manag Bus Excell. 2015;26:11–2.
32. Biasini V. Implementation of a quality management system in a public research centre. Accred Qual Assur. 2012. https://doi.org/10.1007/s00769-012-0936-9.
33. Jefferson T. Quality and value: models of quality control for scientific research. Nature. 2006. https://doi.org/10.1038/nature05031.
34. UK Government: joint code of practice for research (JCoPR). 2015. https://www.gov.uk/government/publications/joint-code-of-practice-for-research-jcopr. Accessed 13 Sep 2017.
35. RQA working party on quality in non-regulated research. Guidelines for quality in non-regulated scientific research booklet. RQA. 2008–2014.
36. RQA: quality systems workbook. 2013. https://www.therqa.com/assets/js/tiny_mce/plugins/filemanager/files/Publications/RQA_Quality_Systems_Workbook.pdf. Accessed 18 Sep 2017.
37. WHO: TDR handbook: quality practices in basic biomedical research. 2010. http://www.who.int/tdr/publications/documents/quality_practices.pdf?ua=1. Accessed 13 Sep 2017.
38. Singapore statement on research integrity. 2010. http://www.singaporestatement.org/index.html. Accessed 13 Sep 2017.

Principles, References and Standards

3

3.1 Quality Basics

The first formalisation of quality as a discipline of operational and strategic management has its roots in the industrial revolution and was reinforced in the military during World War II; then it spread internationally in the manufacturing environments during the 1980s and 1990s. The intuition of Deming and Juran—the undisputed fathers of total quality management—to apply statistical tools and controls not only to the productive activities but also to the organisational processes and management can have widespread application.

The origin and the first applications of quality principles and standards were much tailored to manufacturing, but the business world soon realised the universal value of this formalised approach to management. The use of the techniques and principles rapidly spread from manufacturing to the service industry, forcing a complete revision of the reference standards (ISO 9000:2000) to fit the application to any kind of organisation.

Even though the history of quality, from its inception in the first decades of the 1990's, throughout last century up to today's evolution, and which pervades the entire management philosophy, would facilitate a thorough—therefore better—understanding of the principles, we will not delve into this aspect, inviting the reader to refer to other texts [1, 2].

In the following chapters, we will make a short journey through quality management and its principles, international quality standards and few methodologies and tools whose translation from industry into research will be detailed in Chaps. 5 and 6.

---

[1] CEI 0-12:2002 Approccio per processi e indicatori della qualità per le aziende del settore elettrotecnico ed elettronico.

[2] *Non-fulfillment of a requirement* (ISO 9000): *An observed situation where objective evidence indicates the non-fulfillment of a specified requirement*, US CFR 33, Navigation and Navigable Waters.

© Springer International Publishing AG, part of Springer Nature 2018
A. Lanati, *Quality Management in Scientific Research*,
https://doi.org/10.1007/978-3-319-76750-5_3

3.2   Total Quality Management

Total quality management (TQM) is a set of statistical techniques for managing production and organisational processes. It may seem a naive concept, but there is a great novelty in the application of statistical techniques to organisations, i.e. the use of methods created for operational analysis at a management—therefore strategic—level. TQM considers that both manufacturing and organisational processes can be managed referring to inputs, outputs, measurable performances and related controls. TQM is also known as Company-Wide Quality Control (CWQC), emphasising the shift of quality control from the production of goods to all operating mechanisms in the company.

Total quality management is based on a richer concept of quality: quality is not limited to customer satisfaction and positive market image, but it is also a commitment to the proper use of instrumental and economic resources (efficiency) and attention to productivity (effectiveness). In other words, the two main aspects of quality management are process—how I get there—and product—what I reach. If you consider quality as the analysis and mitigation of possible opportunities for error, then this capacity for prevention is an investment that, using known resources, can avoid the costs of unpredictable errors caused by uncontrolled activities.

3.2.1   The Process Approach

The main structural way to achieve the advantages promised by quality is the process approach. The purpose of the process approach is to enhance an organisation's effectiveness and efficiency in achieving its defined objectives [3]. A process can be defined as a set of interrelated or interacting activities, which transforms inputs into outputs. Figure 3.1 shows the scheme for a generic process and highlights the need to define it in term of inputs, outputs, rules and resources.

The performances of a process can be measured in terms of effectiveness and efficiency, defined as:

EFFECTIVENESS OF PROCESS = Ability to achieve the desired results
EFFICIENCY OF PROCESS = Results achieved vs. resources used

Having defined a mode to gather data, how the process flows and what it can produce, we can control its performance comparing data collected every time the process is replicated.

What then is the difference between a process-oriented organisation and one hierarchically oriented? Organisations are usually managed vertically, with responsibility for the intended outputs being divided among functional units. When an organisation works in a hierarchical manner, there is limited autonomy for operators, and references to those responsible for decision-making are required. Thus operations face a heavy bureaucracy, long lead times and rarely assessed steps; moreover, *the end customer or other interested party is not always visible to all*

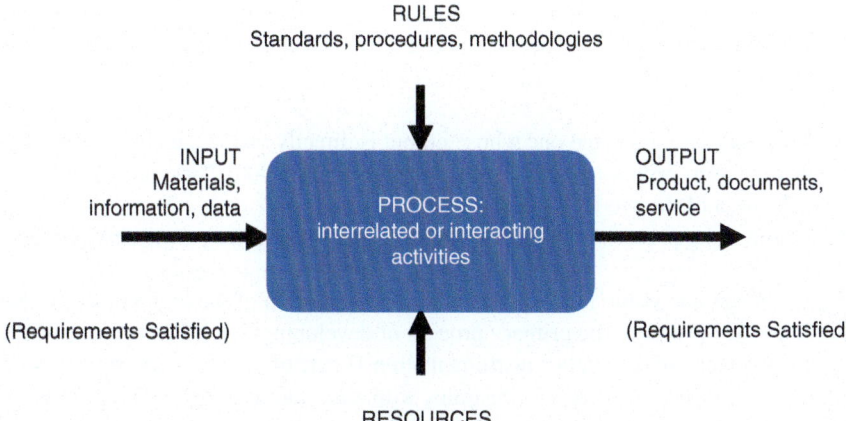

Fig. 3.1 Scheme for a generic process. *Rules* and *resources* influence the *process*, which can be carried out either within the organisation or by an external provider. *Inputs* to the process should be controlled to ensure that they are compliant with the requirements. The same control should be performed on the *outputs*, before releasing them to the final user

involved. Consequently, problems that occur at the interface boundaries are often given less priority than the short-term goals of the units. This leads to little or no improvement to the interested party, as actions are usually focused on the functions, rather than on the intended output [3]. In contrast, in the process approach, the flow of activities is described a priori and execution protocols for each player and each step are defined, along with the time and resources to be dedicated to the task. With the proper empowerment and a coded description of the activities, people have authority for autonomous action and can act without close supervision or the control of another accountable person. They can turn to the higher level only for activities outside the programme or outside normal procedure—this would occur only occasionally.

In the *process approach*, each process is linked to other processes within the organisation in a supplier-customer chain. In defining the output of each process attention has to be paid at the very least, to what, how and when the result is expected by the user. This is a great chance to place the next process owner (the *internal customer*) in the best conditions to work with ease, thus optimising the workflow within the organisation. Ultimately, the organisation's entire activity is driven by customer's requirement; in other words, *the process approach introduces horizontal management, crossing the barriers between different functional units and unifying their focus to the main goals of the organisation* [3].

Two main advantages arise from this view: processes can be measured and thus evaluated, and the description of the organisation as a *system of processes* gives the opportunity to simplify and optimise interfaces between activities, both being great sources of information for improvement.

When the organisation is described by a process map, it is a good idea (following the ISO 9000 approach) to separate them into two classes: *primary* and *supporting* processes:

- A primary process is the one whose output is directly visible and measurable by the final user; primary processes are identified by ISO 9001 Par. 7 as linked to the realisation of the product or service [4].
- A supporting process is a process whose output is utilised by a primary process.

As an example in the field of research, the preparation of the final report for the funding body is part of the primary process of developing a research project, while the maintenance of laboratory instrumentation is part of the supporting process of resource management. Box 3.1 illustrates with a humorous example the difference between the traditional (hierarchical) approach and the process approach.

Box 3.1: Little Red Riding Hood
In order to understand the difference between the approach by function (hierarchical) and the approach by process of an organisation, let's retell the well-known fairy tale of Little Red Riding Hood. According to the hierarchical approach, the story would run something like this:

Little Red Riding Hood: pays a visit to Grandmother, disobeys her mother and goes through the woods, tells the wolf she is going to her Grandmother, reaches Grandmother's home and greets her and thanks the Hunter.

Little Red Riding Hood's Grandmother: frightened, receives the Wolf, and thanks the Hunter.

The Wolf: enquires about what Little Red Riding Hood is going to do, arrives to the Grandma's cottage before Little Red Riding Hood, swallows the Grandmother, disguises himself like the Grandma and waits, and eats Little Red Riding Hood

The Hunter: sees the wolf through the window, enters the house, kills the wolf cutting his belly and frees Little Red Riding Hood and the Grandmother

In such a way, the story becomes unintelligible. When telling the fairy tale to our children, we use instead a process approach: we explain how the story unfolds, we tell the sequence of actions (schedule), describe activities (subprocesses) and depict how each character (person in charge, organisational function) acts.

This curious but meaningful example is drawn from a guideline edited by the Italian Electrotechnical Committee (Comitato Elettrico Italiano, CEI).[1]

Each process has to be performed under the responsibility of a process owner who has to be clearly identified, to avoid misalignments in process management. A good process management can be summarised in five simple steps [3]:

1. Identify key processes.
2. Define quality standards for those processes.
3. Decide how process quality will be measured.
4. Document the approach to achieving the desired quality, as determined by measurements.
5. Evaluate quality and continuously improve.

3.2.2 Principles

TQM identifies a number of principles on which the management of an organisation should be based, should it wishes to use quality as a strategic lever. Since the revision of ISO 9000 in 2000 the standard has integrated the principles of TQM, developing and coding them into eight statements. The 2015 edition has reduced them to seven.

Principle 1: Customer Focus *The primary focus of quality management is to meet customer requirements and to strive to exceed customer expectations.*

It is no coincidence that the first principle of TQM indicates that the survival and success of companies depend on how they are viewed by customers. Successful organisations know how to interpret the latent needs and unexpressed market requirements offering appropriate products. This principle is generally applicable to a company product or service to a public organisation and administration—in this case, attention has to be paid to public opinion.

The next step in focusing attention to the customer is identifying the stakeholders, who is anyone who has an interest in the work and success of the organisation. Stakeholders can therefore be seen as a kind of client to whom the organisation provides products or services that are not immediately perceptible as such. Stakeholders are funders, suppliers, employees, etc. Considering them as customers and orienting efforts towards their requirements, the company takes the opportunity to recruit other forces and other interests around its targets. According to ISO guideline about TQM principles [5], respecting the first principle leads to key benefits such as an *increased customer value; increased customer satisfaction; improved customer loyalty; enhanced repeat business; enhanced reputation of the organisation; expanded customer base and increased revenue and market share.*

Principle 2: Leadership *Leaders at all levels establish unity of purpose and direction and create conditions in which people are engaged in achieving the quality objectives of the organisation.*

TQM refers to a particular style of governance. This principle underlines that leaders should help personnel to understand business objectives, thus creating cohesion of purpose and activities to achieve targets. The new edition of ISO

9001:2015 emphasises this strategic role of high management, consigning to it the responsibility of the quality management system, once delegated to the so-called *management representative for quality*. Following the ISO guideline, the organisation can achieve an *increased effectiveness and efficiency in meeting the organisation's quality objectives; better coordination of the organisation's processes; improved communication between levels and functions of the organisation; development and improvement of the capability of the organisation and its people to deliver desired results* [5].

Principle 3: Engagement of People *It is essential for the organisation that all people are competent, empowered and engaged in delivering value. Competent, empowered and engaged people throughout the organisation enhance its capability to create value.*

The human factor is the primary wealth of any organisation: staff must be involved in policies and business strategies and encouraged to give a personal contribution, within their level of accountability. This goal is accomplished by paying specific attention to the corporate climate and helping employees in their work and in their rightful expectations. Box 3.2 shows a comparison between a statement made by F. W. Taylor (the father of the modern assembly line) and another by K. Matsushita (chief and founder of Panasonic industrial group), about personnel involvement in working and strategic decisions. The third TQM principle is clearly the translation of Matsushita's thoughts. The ISO guidelines show as achievable benefits *improved understanding of the organisation's quality objectives by people in the organisation and increased motivation to achieve them; enhanced involvement of people in improvement activities; enhanced personal development, initiatives and creativity; enhanced people satisfaction; enhanced trust and collaboration throughout the organisation; increased attention to shared values and culture throughout the organisation* [5].

Box 3.2: Taylor's and Matsushita's Thoughts

Hardly a competent workman can be found who does not devote a considerable amount of time to studying just how slowly he can work and still convince his employer that he is going at a good pace. Under our system, a worker is told just what he is to do and how he is to do it. Any improvement he makes upon the orders given to him is fatal to his success.

F. W. Taylor, (1929)

For you, management is the art of smoothly transferring the executives' ideas to the workers' hands. (...) For us, management is the entire workforce's intellectual commitment at the service of the company.

K. Matsushita (1988)

Principle 4: Process Approach *Consistent and predictable results are achieved more effectively and efficiently when activities are understood and managed as interrelated processes that function as a coherent system.*

The former ISO 9000 process approach and system approach principles have been combined into this single revised process approach principle. The process approach allows controlling activities and thus achieving a greater efficiency, both in managing the current assets and in the use of resources. There is an undeniable difficulty in companies embracing the quality and the ISO model, in moving from hierarchical to a process-based logic. Often empowerment—even with regard to issues with no significant impact on the company's structure—is perceived with the anxiety engendered by a lack of control. Conversely, a complete process of suitable instructions and appropriate control points fosters operations and provides management with clear, accessible, comparable information on their trends.

To obtain a complete synergy in the process approach, we have to identify, understand and properly manage their mutual relationships; only then the organisation can collect the resulting benefits in terms of effectiveness and efficiency. In a process-centred organisational scheme, the degree of importance is no longer determined by its hierarchical position but by the critical importance of the process. Accordingly, from the perspective of customer satisfaction, the processes that deal with the customer and their satisfaction inevitably become the primary processes. We can refer to the ISO guideline to find the most important benefits that the fourth TQM principle ensures: *enhanced ability to focus effort on key processes and opportunities for improvement; consistent and predictable outcomes through a system of aligned processes; optimised performance through effective process management, efficient use of resources, and reduced cross-functional barriers; enabling the organisation to provide confidence to interested parties as to its consistency, effectiveness and efficiency* [5].

Principle 5: Improvement *Successful organisations have an ongoing focus on improvement.*

Another distinctive aspect of TQM encourages companies not to just maintain the control of processes, but to focus on and lead to a gradual improvement. This push for improvement should be pursued with conviction by management; otherwise organisations develop a tendency to be satisfied with the results, rest on their laurels and potentially surrender gradually to some lesser objective (see Box 3.3). In this case, the results that may have been obtained with the enthusiasm and commitment in the initial phase could be negated within a short time. As ISO guidelines state, major benefits could be *improved process performance, organisational capabilities and customer satisfaction; enhanced focus on root-cause investigation and determination, followed by prevention and corrective actions; enhanced ability to anticipate and react to internal and external risks and opportunities; enhanced consideration of both incremental and breakthrough improvement; improved use of learning for improvement; enhanced drive for innovation* [5].

Box 3.3: The Frog in the Pot

If you drop a frog in a pot of boiling water, it will hop right out. If you put a frog in a pot of cool water, however, and you gradually turn on the heat, the frog will not notice what is happening. It will stay in the water until it dies.

This little story with a Zen flavour tells us that small, subtle changes can eventually bring major damage without even noticing it. The metaphor, applied to companies and their strategic objectives, teaches us that if we accept small, negative changes to objectives we end up as the unaware boiled frog.

Principle 6: Evidence-Based Decision-Making *Decisions based on the analysis and evaluation of data and information are more likely to produce desired results.*

Many comments and many examples may be given regarding this point. How often technical or strategic decisions are made on the basis of generic information of any kind, even on the feeling, experience or emotional wave of the moment? TQM strongly emphasises (and this is one of the better controlled items during ISO 9001 audits) the importance of quantitative data to describe each situation and to be used in every decision. This principle shows a second, fundamental truth: information can be vital. Being able to pick up weak signals from the field of interest or from a sponsor and to use them to support appropriate strategic decisions can be the key to success for the organisation. Key benefits for ISO guideline are *improved decision-making processes; improved assessment of process performance and ability to achieve objectives; improved operational effectiveness and efficiency; increased ability to review, challenge and change opinions and decisions; increased ability to demonstrate the effectiveness of past decisions* [5].

Principle 7: Relationship Management *For sustained success, organisations manage their relationships with interested parties, such as suppliers.*

If you magnify the process to include all the agents that are part of it, it is immediately obvious that the suppliers of a company will significantly influence the end beneficiary with their products and/or their services. A relationship between customer and supplier based on honesty, trust and collaboration, by establishing a strong partnership, fosters the development and the sharing of common goals. With respect to the previous version of this principle (ISO 9000:2000), the new ISO 9001, in the clause 4.2 [4], identifies possible interested parties as direct customers, end users, suppliers, distributors, retailers and regulators. Real advantages from the application of this principle are, according to ISO guideline, *enhanced performance of the organisation and its interested parties through responding to the opportunities and constraints related to each interested party; common understanding of goals and values among interested parties; increased capability to create value for interested parties by sharing resources and competence and managing quality-related risks; a well-managed supply chain that provides a stable flow of goods and services* [5].

3.2.3 PDCA Cycle

The PDCA cycle (Fig. 3.2) was originally introduced by the American scholar of quality, Walter Shewhart in the 1920s and 1930s, later taking the name of *Deming cycle*, as it became the cornerstone of the discipline he founded. Because of its importance and its very broad application in the field of quality, PDCA is considered the main concept of TQM and subsequently of ISO 9000 standards. It is a logical approach to problem solving, used to maintain performance and to identify and pursue opportunities for improvement. It consists of four simple, iterative steps: plan, do, check, act, from which the acronym PDCA.

PLAN:

- Identify the problem or opportunity, analyse, find the solution and plan the intervention.

DO:

- Run the planned intervention and collect data.

CHECK:

- Review the results and compare with the objective, and identify what has been learnt.

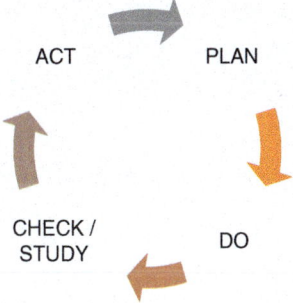

Fig. 3.2 PDCA cycle phases are *Plan* (define goals/targets and evaluation criteria, schedule actions), *Do* (execute action plan), *Check* (evaluate results with respect to targets/goals) and *Act* (consolidate positive results and good practices, or define a new action plan if results are not satisfactory)

ACT:

- Repeat the process to refine the results; if positive, incorporate what has been learnt into procedures to consolidate the experience.

Iteration of the process is necessary when the objective is not reached. Data collected in the CHECK step will be part of further PLAN analysis and definitions. Even if the goal has already been reached, iterating allows refining the understanding and control of the process, thus allowing improvement.

Late in his career, Edward Deming renamed it Plan Do Study Act (PDSA), contending that *Check* would emphasise the control at the expense of the understanding of the process required by that stage. The change intended to underline the use of existing knowledge to better understand the product or process being improved; observing and learning from the consequences (study)—after checking the result (check)—lead to determining what modifications should be made (act).

PDCA can also be used to break down a project into different phases (useful in a working breakdown structure—WBS, see Chap. 5 for project management tools) and related planning. See an example in Box 3.4.

Box 3.4: A Plan Based on the PDCA Cycle for Implementation of a Quality Management System (QMS)

Here's how an action plan for setting up a quality management system (QMS) can be structured:

PHASE I (PLAN)
- Analyse and assess the organisation.
- Define a scheme of main processes.
- Plan the QMS implementation.

PHASE II (DO)
- Give support to any organisational changes.
- Design the QMS (creation of a system based on procedures).

PHASE III (CHECK)
- Implement the QMS (verification by application).
- Audits.
- Management review.

PHASE IV (ACT)
- Make changes identified by testing and verification.
- Consolidate the QMS.

3.2.4   The Requirements

The concept of requirement was introduced by the radical revision of the 2000 standard and is defined as *the need or expectation that is stated, generally implied or obligatory*. The principle of a quality management system is all about requirements—namely the ability of an organisation to incorporate the expectations of customers and turn them into product specifications: thus meshing the requisites of the standard and those of customers, with the sole purpose of running the organisation effectively and efficiently, and therefore achieving the satisfaction of the customer and all stakeholders.

We can identify three kinds of requirement:

- Explicit/stated: declared in contracts and transactions with customers, for example, technical characteristics of the product, service conditions, agreed defect rate
- Implicit/implied: not declared because taken for granted, for example, compliance to laws and rules
- Latent: undeclared and often unconscious expectations, what can be expressed with *it would be nice if …*

In a research project, for example, explicit requirements are the scope, the goals, the schedule and the deliverable. Implicit requirements can be the good conduct of the research, the proper treatment of animals and the respect of national laws. A work package for the coordination of the quality aspects of the project could be a latent requirement for the funding body. Particular attention has to be paid to latent requirements, because once declared they become implicit, and thus to be taken into account in future cases. To make an analogy, nowadays everyone expects a camera embedded in the mobile phone, while a few years ago the first cellular phone producer who offered it impressed the market capitalising on a latent expectation.

3.3   Quality Certification and Accreditation

By *certification* we mean the act by which a certification body, accredited nationally or internationally, declares that a product, process, service or business system complies with a specific standard. The certification process should be seen not as an inspection, but as a measure of conformity to a reference standard. Agencies involved in the management of certification are:

- ISO International Organization for Standardization: defines the standards and related time frames for upgrading
- National Accreditation Body: for each country, accredits certification bodies in accordance with ISO/IEC 17021:2006, i.e. verifies their requirements and recognises them as adequate

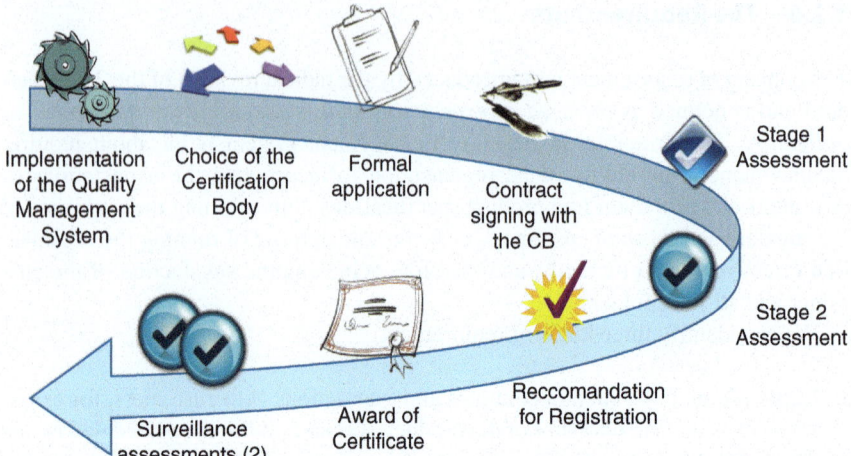

Implementation   Choice of the   Formal
of the Quality   Certification   application
Management   Body   Contract
System   signing with
                                                the CB

Stage 1
Assessment

Stage 2
Assessment

Surveillance   Award of   Reccomandation
assessments (2)   Certificate   for Registration

Fig. 3.3 Certification process for a generic organisation. The organisation willing to achieve the
ISO 9001 certification shall implement the QMS, choose the certification body and sign with them
a standard contract. Then the organisation is evaluated in two stages and upon a positive result it is
recommended for certification; this is issued after the deliberation of the decision-making body of
the certification authority. Surveillance assessments are executed for the next 2 years to complete
the 3-year cycle

• Certification bodies: certify the company quality systems—that is, verify their
  compliance with respect to the standards

Accreditation differs from *certification* by adding the acknowledgment, made by
an accreditation body, of the technical competence specific to a scope of accreditation, in addition to its compliance to a documented quality system.

Figure 3.3 shows the certification process for a generic organisation. After the
organisation has implemented a quality management system (QMS) and signed a
contract, the certification body (CB) performs an analysis of the main documents
of the QMS (namely the quality manual, the management review report and the
internal audit report) and a site visit. This step is named Stage 1. Then a second,
complete audit named Stage 2 is conducted to verify the correct application of the
standard requirements and the internal rules described in the QMS. If no non-
conformities[2] are detected, the organisation is proposed for registration and the
CB, after having analysed the audit report, issues the certificate. In case of minor
non-conformities, the organisation is certified, but it must eliminate the causes of
non-conformities by means of adequate corrective actions, to be realised in due
time before the next CB audit. One or more major non-conformities, i.e. not
proper standard implementations that jeopardise the quality approach, preclude
the certification. In that case, the organisation must issue an improvement plan
and realise the corrective actions within a short time, to allow a new certification
audit by the CB.

Two more surveillance audits are made in the next 2 years by the CB to assess the actual application and compliance of the QMS.

Attention should be paid to the difference between product certification and system certification: the former is related to a single product, attesting its conformity to the requirements established by technical rules and/or technical standards or equivalent documents. The latter attests the conformity to a standard of the entire company's management system, which is recognised as able to provide conforming products and services.

In any case, the certification of an organisation does not automatically guarantee the quality of products or services provided; but it is evidence that the organisation management is consistent with the principles of quality; this, in turn, should ensure the quality of all the organisation's products—materials and/or services—from their inception.

The adoption of a QMS integrating the traditional management system of the organisation acts as follows:

- Keeps the focus on the company's ability to meet customer requirements.
- Spreads within the organisation the culture of performance monitoring (objectives, indicators, monitoring, checking).
- Provides regulatory and operational tools to involve all staff in the achievement of business objectives, including customer's and stakeholders' satisfaction.
- Requires a critical analysis of the organisation's modus operandi, seeking out weaknesses and inefficiencies.
- Creates the conditions for improvement.
- Endows the company with the potential for competitive advantage.

Applying quality principles and the ISO 9001 standard model to the organisation brings clear advantages in terms of efficiency and effectiveness, which are independent from the certification of the management system. Manders and de Vries [6] outline the advantages of having a quality management system according to ISO 9001 and having the related certification, in terms of internal, external and image benefits, as illustrated in Fig. 3.4. As it can easily be seen, the main internal benefits stand on the increased ability to control the scientific process, the improved quality of scientific results and the higher scientific productivity, all contributing to effectiveness and efficiency. Moreover, better internal results favour external expansion, allowing access to better funding opportunities, at the same time improving the related success rate. The certification itself acts on the image and reputation of the research institution, fostering the attraction of funding and allowing a larger cooperation with other institutions, as well as new opportunities if the research unit is offering work on order. Above all, together with an increased ability to attract funds, the research institution can *do better with less.*

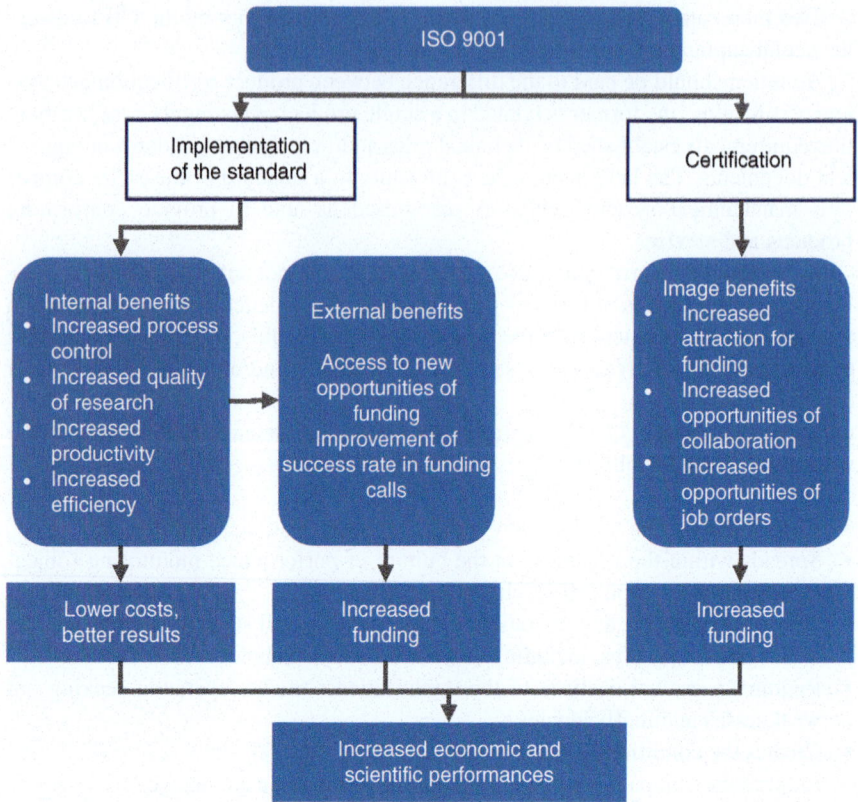

Fig. 3.4 Advantages of a QMS based on ISO 9001 standard for scientific organisations. The implementation of the standard gives internal and external benefits, while the certification gives only image (external) benefits. Overall, the application of the ISO 9001 standard generates both economic and scientific advantages (adapted from [6])

3.4   Quality Standards for Certification in Scientific Research

Although to date there are no international quality standards dedicated to basic scientific research, a scientific laboratory can take as a reference one of the following:

- ISO 9000 is a set of standard intended for a generic quality management system. It originates from the industrial field and it has been generalised for use in any organisation. An organisation can ask for a certification that will be based on the ISO 9001 *quality management system—requirements* standard [4]. There are very few examples of scientific laboratories certified with respect to this standard.

- Good laboratory practice (GLP) is mandatory in most countries for scientific preclinical research devoted to the development of human and veterinarian drugs. Outside this field, it can be kept as reference but no certification nor accreditation is envisaged.
- ISO/IEC 17025 is the reference for testing and calibration laboratories and it leads to accreditation; it is useful for a lab providing quantitative analysis for external organisations, thus taking advantage of accreditation, but it could be an excessive burden of management for a research lab.

All of these standards provide useful guidance on how to organise laboratory quality management: while ISO 9000 and 17025 devote attention on the management system, GLP focuses on research practices, requiring strict controls and records. ISO 17025 itself is mostly focused on ensuring the accuracy of measurement and instruments. In the following paragraphs, information is given for those who want to improve at least one of the aspects mentioned. To find directions dedicated to scientific research, even when not certifiable, please proceed to Chap. 3.

3.4.1   ISO 9000

The ISO—International Organization for Standardization—is an international organisation with headquarters in Geneva (CH), which is responsible for defining and managing standards. Its goal, according to its constitution (dated 1947), is *to facilitate the international coordination and unification of industrial standards*. The ISO 9000 family of standards defines and regulates corporate organisations and their operation, providing a reference model to achieve the necessary quality of products or services. It was issued for the first time in 1987 to relieve the companies from the need of auditing their suppliers. This activity would be needed to ensure that suppliers have applied a quality approach. ISO 9000 standards are derived from US military standards (MIL-STD) and UK-Standard; in the first draft they were mainly focused on prevention, catalysed by quality assurance.

The second edition of 1994 was updated to take into account market trends, without changing the standard structure: it was still oriented to manufacturing and, because of its excessive attention to formalising and documenting actions and decisions, was likely to induce a so-called *ISO bureaucracy*. The third edition of 2000 has seen a complete renewal of the structure and content of the standard, in order to extend its use to the production of services. This edition emphasised the concepts of process management, continuous improvement of processes and customer satisfaction, which shall be measured and monitored, as well as the requirement for a documented system (thus not *system of documents*). The edition of 2008 did not make any substantive changes, while the 2015 edition has brought great innovation in the approach to the management system, introducing risk-based thinking, valuing the role of leadership and streamlining the use of formal documentation.

The ISO 9000 family of standards is composed of three documents for establishing and managing the quality system, plus a reference for the audits. Since the global issue of 2000, each single document has undergone independent updates:

- ISO 9000:2015, quality management systems—fundamentals and vocabulary.
- ISO 9001:2015, quality management systems—requirements.
- ISO 9001:2009, managing for the sustained success of an organisation—a quality management approach
- ISO 19011:2011, guidelines for auditing management systems

The first document (ISO 9000:2015) is dedicated to illustrating general aspects and providing a common vocabulary. ISO 9001:2015 is the reference for the certification of a quality management system, and describes organisation by processes and the requirements for the QMS to be compliant to the model. ISO 9004:2009 completes the family: it provides guidance for organisations that, having implemented a quality system, wish to exploit its full potential for improvement; this can also be applied independently and then it is not intended for certification nor contractual purposes.

3.4.2   Good Laboratory Practice

First issued by the Food and Drug Administration in 1978, the good laboratory practice (GLP) has the aim to regulate *how to perform studies in a scientific and ethical manner and how to record the data obtained from the studies* (FDA). Basing itself on this first document, the Organisation for Economic Cooperation and Development—*OECD*—set a working group to generate an international standard. Participants were Australia, Austria, Belgium, Canada, Denmark, France, the Federal Republic of Germany, Greece, Italy, Japan, the Netherlands, New Zealand, Norway, Sweden, Switzerland, the UK, the United States, the Commission of the European Communities, the World Health Organisation and ISO, agreeing on a suitable text for a suggested common use in 1981. The OECD GLP is the only guideline that is accepted universally, since it has been discussed by an international panel of experts and has been agreed on at an international level. FDA (Food and Drug Administration, USA) and other institutions (e.g. US Environmental Protection Agency, EPA) have adopted slightly different standard texts, usually tailored to the specific applications involved. On the FDA website [7] a comparison between OECD, EPA and FDA GLP can be found.

GLP applies to test facilities carrying out preclinical studies on human and veterinarian drugs, assuring the quality and integrity of data submitted to regulatory agencies, in support of the safety of regulated products. As per OECD's definition, GLP is a quality system concerned with the organisational process and the conditions under which non-clinical health and environmental safety studies are planned, performed, monitored, recorded, archived and reported. With respect to the ISO 9001 model, this means that GLP is not a complete quality management system, as it deals only with the main process of research.

To easily remember the core of GLP's principles, there are some simple rules that can be synthesised as *Say what you do, do what you say, prove it and improve it.* In other words, write standard operating procedures (SOP), follow the procedures, keep good records of what you are doing and continue to improve. SOPs are the description of any operation performed during the research programme and their execution must be documented (i.e. registered). Great attention is paid to data and especially raw data, which must be registered in the original, without omissions, in indelible ink, corrected if necessary in a legible way and signed. The entire study must be documented in order to be easily reconstructed, with respect to SOPs, equipment, materials and data: GLP gives rules and suggestions to facilitate the drafting of the required documents.

GLP gives also special attention to other aspects crucial for a good research process: i.e. description of the role of study director, quality assurance and other organisational figures, training of personnel, adequate and clean facilities, and well-maintained and calibrated equipment.

3.4.3   ISO/IEC 17025

ISO IEC 17025, *General requirements for the competence of testing and calibration laboratories*, is a technical standard for accreditation: it needs a formal observance to ISO 9001 organisational requirements and further specific technical requirements. In such a way, the accreditation can ensure that a laboratory supplies reliable and accurate test results and measurements.

ISO/IEC 17025:2005 [8] is intended for laboratories performing tests and/or calibrations, including sampling. The standard gives general requirements for testing and calibration performed using standard methods, non-standard methods and laboratory-developed methods. The organisations that may apply this standard include, for example, first-, second- and third-party laboratories, and laboratories where testing and/or calibration are carried out within the frame of inspections and product certifications.

When the requirements of products or services may have an impact on public health and safety, the standard for traditional management systems is integrated with a specific standard, which includes more stringent and distinct obligations and requirements. ISO 17025 adds, to the general requirements listed in ISO 9001, organisational and technical requirements dedicated to testing laboratories. Among the organisational requirements are a management system with documentation, improvement, audit, non-conformities and complaint process management, as well as more detailed recommendations about contract review, service outsourcing, customer service and procurement. Technical requirements are linked to what determines the correctness and reliability of the tests and calibrations performed in the laboratory, such as identification and maintenance of the equipment, traceability of measurements (reference standards and reference materials), sampling, handling of test and calibration items, reports and certificates.

3.4.4   Other References

There are a few additional standards in the field of life science and healthcare, which could be useful as a reference for specific research issues. In the following paragraphs two examples are described in broad outline.

ISO 15189, *Medical laboratories—requirements for quality and competence*, is used by medical laboratories in developing their quality management systems, with respect to peculiar aspects, and assessing their own competence, with regard to customers, regulatory agencies and accreditation bodies. ISO 15189 adds specific requirements to ISO 9001 and 17025, taking into consideration few specific management processes, such as:

* Interpretation of the results of analysis and formulation of the response
* Logistics (including services related to patient care) and patient environment
* Reception of patients
* Collection of samples from patients
* Ensuring that the procedures used are able to meet the requirements of the contract and the clinical needs
* Advice regarding test selection and use of services, including different types of samples required and frequency of their repetition

ISO 15189 also includes ensuring laboratory competence (e.g. personnel, facility, instrument and examination methods), participation in proficiency testing, traceability of measurements and control of uncertainty of measurement. This standard is applicable, for example, to any laboratory performing genetic or clinical analysis.

ISO 13485, *Medical devices—quality management systems—requirements for regulatory purposes*, is a voluntary standard for the certification of quality management systems as applied to the design and manufacture of medical devices. ISO 13485 aims to facilitate implementing a quality management system dedicated to medical devices taking into consideration *harmonised regulatory requirements* [9]. The organisations that only apply the ISO 13485 standard may exclude some of the requirements of ISO 9001 that are not applicable as regulatory requirements to the field of medical devices, and thus they cannot claim conformity to ISO 9001.

References

1. Lanati A. Qualità in Biotech e pharma: gestione manageriale dalla ricerca ai suoi prodotti. Springer; 2009.
2. Kiran DR. Total quality management. 1st ed. Oxford: Butterworth-Heinemann; 2016.
3. ISO: guidance on the concept and use of the process approach for management systems, ISO/ TC 176/SC 2/N 544R3. 2008. https://www.iso.org/files/live/sites/isoorg/files/archive/pdf/ en/04_concept_and_use_of_the_process_approach_for_management_systems.pdf. Accessed 13 Sept 2017.
4. ISO 9001:2015 quality management systems—requirements.

5. ISO: quality management principles. 2015. https://www.iso.org/files/live/sites/isoorg/files/archive/pdf/en/qmp_2012.pdf. Accessed 13 Sept 2017.
6. Manders B, de Vries HJ. Does ISO 9001 pay?—analysis of 42 studies. 2012. https://www.iso.org/news/2012/10/Ref1665.html. Accessed 13 Sept 2017.
7. US Food and Drug Administration: comparison chart of FDA and EPA good laboratory practice (GLP) regulations and the OECD principles of GLP. 2004. http://www.fda.gov/iceci/enforcementactions/bioresearchmonitoring/ucm135197.htm. Accessed 13 Sept 2017.
8. ISO: ISO/IEC 17025:2005 general requirements for the competence of testing and calibration laboratories. 2005. https://www.iso.org/standard/39883.html. Accessed 13 Sept 2017.
9. ISO: ISO 13485:2003 medical devices—quality management systems—requirements for regulatory purposes. 2003. https://www.iso.org/standard/36786.html. Accessed 13 Sept 2017.

Quality and Basic Biomedical Research

4

4.1 What Is Basic Biomedical Research

While preclinical research and clinical research have their own regulatory references—respectively good laboratory practice (GLP) and good clinical practice
(GCP)—there is a wide research area left to the competence, the rigour and the
intellectual honesty of the scientists who pursue the scientific method. According to
the WHO-TDR handbook for Quality Practice in Basic Biomedical Research—
QPBR,[1] the entire flow of research, starting from the first idea to its exploitation, can
be outlined as in Fig. 4.1. From this scheme it can be seen that basic biomedical
research together with CMC (chemistry, manufacturing and control) research are
the two areas without a regulatory standard. The QPBR defines basic biomedical
research as *activities intended to find means of detecting, preventing or treating
human disease. Such research covers the discovery and exploratory studies that
precede the regulated phases of drug development or programmes to develop other
methods of disease control* [1].

More specifically, the QPBR distinguishes three phases of basic biomedical
research, as shown in Fig. 4.2, for both drug development and basic research. In
both cases, the research starts from literature or observations; it develops into a
specific hypothesis with its related test plan, and ends with the proof, based on
experimentation and observation, of the principle identified. Sometimes, basic biomedical research is referred to as *non-regulated research*, even though, more precisely, only one part of this research is not regulated.

---

[1] See Sect. 4.4.1.

© Springer International Publishing AG, part of Springer Nature 2018   43
A. Lanati, *Quality Management in Scientific Research*,
https://doi.org/10.1007/978-3-319-76750-5_4

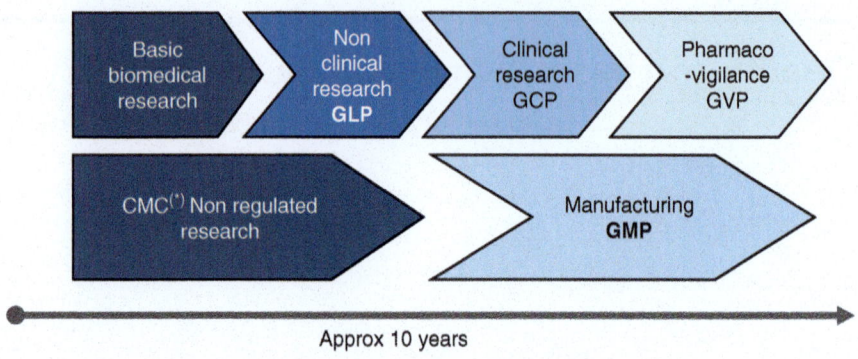

Approx 10 years

(*) CMC = Chemistry, Manufacturing and Control

Fig. 4.1 The entire body of biomedical research. Non-clinical research, clinical research, pharmacovigilance and manufacturing are governed by GxP. Basic biomedical research and CMC are highlighted as both lack international quality standards (adapted from [1])

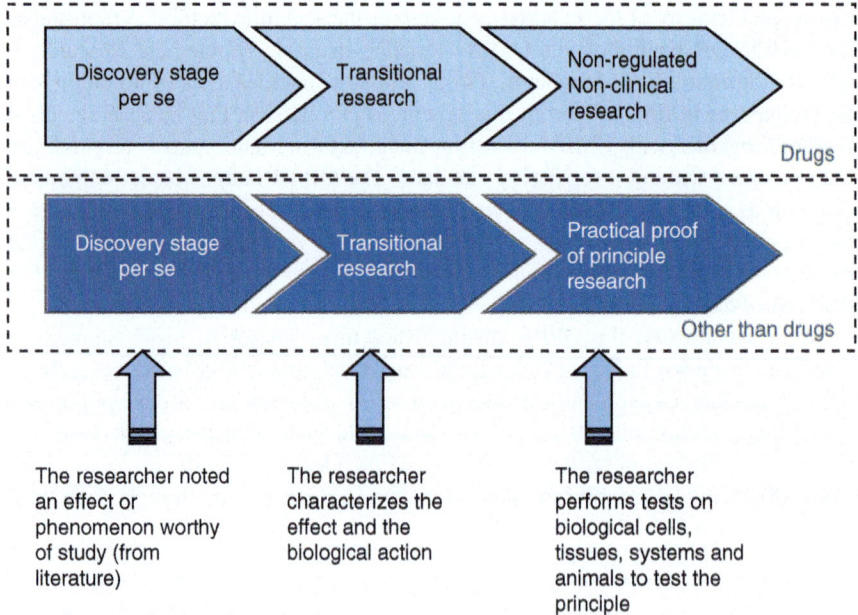

Fig. 4.2 Phases of basic biomedical research. The parallel streams for research on drugs and products other than drugs show the same three-stage evolution: discovery, characterisation and testing (adapted from [1])

4.2 Conscious Management

Results of this first stage of research will be the foundations for the development of useful products and medical protocols for fighting diseases. In the international scientific world, the work of many thousands of researchers will be based upon the results achieved by those who have worked at the basic, initial phase. Thus, their results have to be solid enough to allow a good and useful deployment, so as not to cause wastage of resources, time and money, as a result of ill-formulated hypothesis; in other words, they have to be *reliable*.

Reliability is achieved by taking care of two fundamental aspects: (1) starting with a good idea, and (2) good research conduct. Without one or the other, it is not possible to produce good data that will constitute a sound starting point for further developments. While the former is a matter of scientific knowledge and intuition, the latter is a matter of management. In the Joint Code of Practice for Research (JCoPR[2]), this same concept is expressed in the *Principles Behind the Code*:

- The quality of the research process (QP)
- The quality of the science (QS)

where QS refers to the scientific project and the *extraction of new knowledge and understandings*, while QP refers to *processes and procedures used to gather and interpret the results of the research*. The final statement of the JCoPR is that *without QP, confidence in the research findings is much reduced*.

What do scientists need? To be aware of the entire management process of scientific projects, what we call *conscious management*. In other words, they have to be prepared to understand and govern the process of scientific research (see Fig. 4.3): understanding requirements, defining targets, planning projects, controlling documents, materials and instrumentation, efficiently managing resources, preparing adequate outputs to technological transfer and, last but surely not least, demonstrating reliability to funding bodies. Management would seem a very difficult challenge, causing resources to be diverted from scientific research, unless a well-defined and tested model can be followed.

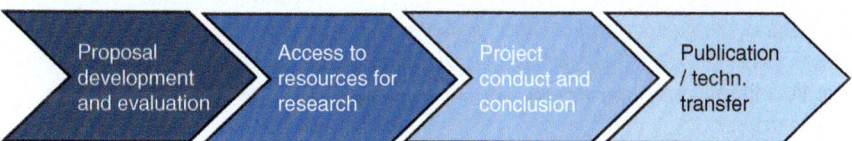

Fig. 4.3 Macro-process of biomedical research, from discovery to publication. The final result of the research, namely a publication, a technological transfer or a patent, is achieved starting from the development and evaluation of a proposal, collecting the required resources, conducting the study and drawing the conclusions

---

[2] See Sect. 4.4 below.

Quality, in its wider meaning, can help on two fronts: on the one hand, it provides a model of organisation, which should facilitate the management aspects, helping to create a favourable environment and to release resources allocated to management for the actual research; on the other hand, quality also provides technical tools and organisational support to the development, verification and transfer of research products. For this latter case, we will see some useful tools in Chap. 4. Focusing the attention on organisational matters, quality can give a significant support to different areas: thinking by processes in a holistic approach encourages to consider the entire path from the very first idea to the final product. In that way, attention is focused on the main process, as seen in Fig. 4.3, and to identify input and output of each sub-process, from initial requirements to final target. This is also the starting point for any research project, allowing identification of the support processes that are needed in order to maintain the correct running of the main process: management of documentation, resources, personnel, materials and instrumentation. As seen in Chap. 2—Total Quality Management—one of the main advantages of the process approach is the opportunity to easily monitor processes establishing verification points, indicators and a control strategy. As an example, every funded research programme requires control milestones and a planned delivery of results. Quality helps the principal investigator not only to set these milestones in a proper way, but also to identify some metrics and the timing (when, how frequently) to check them: planned metric checks can tell the correct development of the project towards the objectives of time and expenses (control of the process), as well as of results (control of the product).

As stated in the third TQM principle, people are a key resource for the organisation, so it is important to exploit and develop their competence and ability as related to the organisation: communication, leadership and team work. This is of particular importance in an environment where, until recently, scientists used to work alone or in small groups, and so they were not accustomed to managing the social aspects of broader collaboration nor the coordination of teams, as it occurs in large and international companies. Some useful hints will be given in Chap. 4 of this book.

4.3   Quality Throughout the Research Project

To better understand the following concepts, we now have to define precisely the difference between *project* and *process*. The Project Management Body of Knowledge (PMBOK) [2] defines a *project* as a temporary endeavour undertaken to create a unique product, service or result. We can then expand:

A project:

- Has a definite beginning and a definite end date
- Has a customer/final user
- Creates something different from the existing
- Allows changes to scope, schedule or requirements, if agreed with sponsor, stakeholders and/or final user
- Follows a life cycle that can be defined as initiation, definition, planning, execution and closure

As we have seen in Chap. 2—Total Quality Management, the Process Approach—a process is a repetitive collection of interrelated tasks, aimed at achieving a certain goal. In more detail:

A process:

- Is ongoing with no clearly defined beginning and end points
- Once validated, it is repeatable and repeated
- Does not allow for changes: a change to a process can only be introduced following a modification project
- It is controllable at checkpoints

Both project and process must be compliant with a set of rules and subjected to checks, in order to guarantee the results to be reliable (therefore effective), and to guarantee the best use of resources (therefore efficient). When researchers begin a study, they are managing a project. The entire project is developed following several processes from the definition of requirements to the exploitation of results. Figure 4.4 shows a model for a project management in research [3], based on the PDCA circle. A more detailed analysis highlights the key interaction with the sponsor, who acts as the *customer* of the ISO 9000 model. Technically, a sponsor is a figure with overall accountability for the project. In a research environment, the sponsor may be, for example, the director of the study, the director of the research institution or the

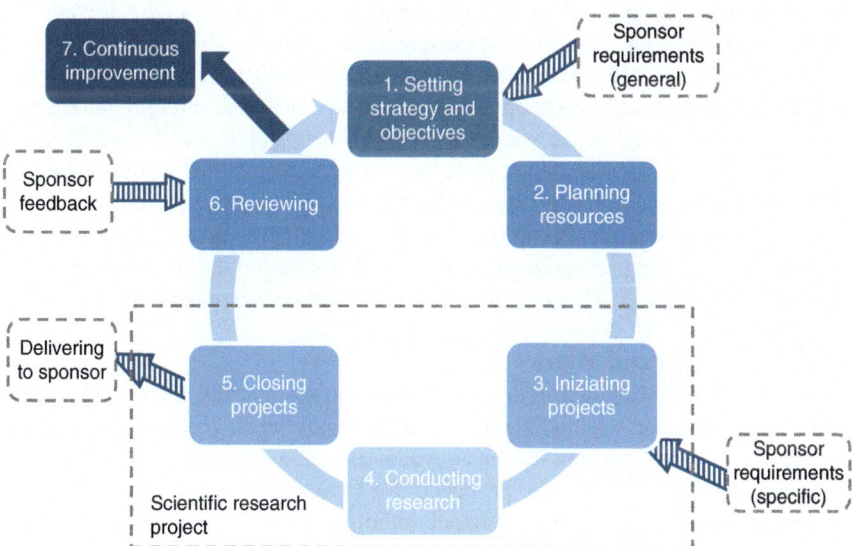

Fig. 4.4 Model for project management in research. The cycle starts with *setting strategy and objectives* that must comply with sponsor's requirements, and *planning resources adequately*. The research project is split into three phases: *initiating*, in which the requirements are detailed and transformed in a project plan; then *conducting the research* and *closing the project* allow *delivering* results to the sponsor. From a quality point of view, it is of paramount importance *reviewing* what has been done and to consider the *sponsor feedback*, to address *continuous improvement* (adapted from [3])

funding body. The owner of the research process first has to define with the sponsor the general requirements, and then—once the project starts—the specific ones. After the development of the project, he/she delivers the results to the sponsor and collects feedback from the sponsor. This feedback is integrated with other results, indicators and controls gathered during the project and at its conclusion, with a view to substantiate periodic reviews and generating suggestions for improvement. As requested by the PDCA model, the circle is then closed with a refining of strategies and objectives. More specifically, activities 1–2 relate to the PLAN phase, 3–4–5 to the DO phase, 6 is CHECK and 7 is ACT. As Robins states, *the processes are generic, and applicable to any type of research activity: basic, strategic or applied.*

In the following section, we are going to see how quality provides references and models, for both management of project/organisation and experimentation.

4.3.1   Management of Study Contents and Controls

Throughout the project, we may need some support in managing general issues as initial project requirements, indicators, documentation, training, communication, procurement as well as good methodologies to streamline trials and analysis of results.

4.3.1.1 Initial Requirements

Usually a research project starts with an idea that rarely foresees its actual final result and, even less frequently, its application in industrial terms. In any case, it would be wise to make clear the objective and the quality level that is expected from the final research results, in order to keep the project development consistent with the target. A quality system gives a clear directive on how to clarify the initial requirements, review them and make the working team aware of them and of how they relate to the final target. In this way, the study can be focused in the best direction, thus obtaining the best from the work of the team. If the scientists involved in basic research see potential applications for their project, they should identify and keep in mind the constraints that industrial research must observe to develop the product. See more details in the section dedicated to project management and exploitation.

4.3.1.2 Controls and Metrics

One of the basics of a quality system is the need to set controls, i.e. checkpoints at which significant data are measured in, in order to govern the processes. In a scientific project, we can measure organisational aspects and scientific aspects, in terms of both process (how I work) and product (what I achieve). Organisational processes are, for example, procurement, personnel training, documentation management and communication. A general principle suggests defining at least one *process indicator* and one *product indicator* for each main organisational process. Table 4.1 proposes some process and product indicators.

When the economic aspect becomes dominant, it is imperative to have the correct perception of the actual costs, the use of resources and the quality of the results produced, not to mention the usefulness of such awareness in a granted research project, when—often scarce—funds shall be managed, at the best, to reach the declared goal. Moreover, a research centre, regardless of its legal nature, type and size, may *sell* competencies, research activities or support to research in an industrial setting especially for the industrial and commercial development of the results of scientific research. An organisation managed by processes, as indicated by the ISO standards, is able to readily calculate costs, having standardised activities, having registered an outline of related activities and being able to assign to each an economic value. This economic awareness is particularly useful in the case of industry-sponsored projects, which seldom give funding, but more often reimburse expenses, including staff salaries. The proper organisation of a research centre might then have a heightened awareness when formulating requests and generating the final balance.

How to control the management of a specific study? There is the need to measure that the performance of the project is as consistent and productive as expected. Throughout the development of the study, we must refer to the three criteria of project management—time–cost–quality—and verify whether the project has remained in line with expectations (usually registered in the funding contract). Finally, the assessment of the value of a study is done mainly through the impact factor (IF) of the publications as issued and accepted by the scientific community. This parameter gives an indicative value tied to the average value of the journal and, by simple transposition, a measure of the scientific value of the study. Table 4.2 shows a few examples of indicators for the management of scientific work.

Table 4.1 Process and product indicators for the organisational processes

Organisational process	Process indicator	Product indicator
Procurement	Percentage of purchase requests processed on time	Percentage of purchase requests processed correctly
Document management	Average time needed to release procedures	Number of errors found in issued documents
Personnel training	Respect of training plan (delays)	Number of non-conformities related to staff skill

For the three examples of the organisational process, one process and one product (outcome) indicator are suggested

Table 4.2 Process and product indicators for the scientific research processes

Scientific processes	Process indicator	Product indicator
Study management	Number of working hours dedicated to single study with respect to plan	Respect of final budget
	% of delay with respect to milestones/deliverables	% of work package targets correctly achieved
Scientific management	% of failure/rework of trials	Number of publications
	Number of external collaborations/reviews/relationships	Importance of publication (impact factor)

The two process and product (outcome) indicators are suggested for the two typical scientific processes: study management and scientific management

Indicators for the quality of research are not so easy to define, yet this assessment is of great importance for the funding bodies when deciding the allocation of funding. Usually those studies and those research teams that put forward values consistent with the principles of the funding bodies are rewarded and supported. In Box 4.1 is an example of how a major funding body evaluates the quality of scientific projects. In conclusion, the main evaluation parameters (the most important characteristics of the scientific study), as well as related evaluation criteria (how to express an assessment), must be identified. The difficulty in this case is that the evaluation of the research cannot be done on predefined standards, but is entrusted to the expert's judgment, personal experience and skills. The issue then shifts from the evaluation criteria, difficult to be encoded, to the choice of experts entrusted with the evaluation, and to whom must be given unconditional trust. From a quality point of view, control of the evaluation is achieved through a careful choice of the *supplier* of assessments, and the careful definition of the procedures for the assessment phase. If the researchers are aware of the evaluation model, they can better sketch the project and, throughout the course of the funded research, will be aware of the aspects that are to be supervised to achieve satisfactory results.

In conclusion, the setup of process control and the analysis of the indicators should enable the research to approach the choices strategically, and design new research lines according to criteria of best outcome, in order to enhance not only the scientific importance of the research but also the ability to obtain funding.

Box 4.1: The Criteria for the Evaluation of Science

Assessing the value of scientific research is crucial for the validation of research results to be published in scientific journals, as well as for selecting research projects to be funded. In both cases, peer review, i.e. the competent and independent evaluation by peers, is internationally acknowledged as best practice. Fondazione Telethon, an Italian charity supporting research on rare genetic diseases, applies peer review principles and procedures to all of its funding decisions for both competitive calls for extramural projects and intramural initiatives. Fondazione Telethon's peer review process relies on an international Scientific Committee made up by about 30 members and on external reviewers, all working abroad. Besides assessing the scientific merit of the proposals, according to established criteria such as the significance, originality and innovation of the project, the appropriateness of the experimental plan and the standing of the applicant in the field, reviewers are asked to score the *impact on patients*, an indicator of how close the expected results are to providing a benefit to people affected by rare genetic diseases. A professional office of former researchers with prolonged experience in the biomedical field manages the entire process, and provides the necessary separation between those who judge and those whose applications are evaluated. The Telethon peer review process is quality certified according to ISO9001 standards since 1994.

Fondazione Telethon also monitors the outputs and outcomes of the funded research. Scientific publications ensuing from Telethon grants are tracked and national and international comparisons are performed by applying bibliometric indicators based on article citation numbers. More importantly, the transition of research from basic studies to translational and clinical activities is closely monitored and facilitated, to fulfil the commitment to find solutions for genetic diseases.

Lucia Monaco, Chief Scientific Officer
Telethon Italy

4.3.1.3 Documentation

Documentation is essential to projects. We are accustomed to consider it mainly for publication and financial and scientific reporting, but it is more than this.

The habit of not fully and exhaustively record and not to register experimental conditions and intermediate results of trials is most likely to cause the loss of important information or references. The researcher is well accustomed to condense the work of years in publications and articles, where only the trace of what led to positive results is recorded. It is rare for the researcher to keep traces of the negative results or those not used in the context of the publication, but all information—if recorded—is very useful, e.g. to avoid following wrong pathways in the future, as a source of inspiration for different projects or to explore alternative approaches. A quality system helps to define document categories, flow of information and essential rules to effectively record, and efficiently store and retrieve information. This same need for information under a quality system also has to be met for the materials used and for the experimental procedures followed. The researcher should develop the habit to document the products used—type, quantity, supplier and internal production methods—as well as the experimental procedures—better if standardised, and, if at all possible, also recording exception and deviations from the standard. This conduct will strengthen the reproducibility of outcomes and facilitate the validation of results and the drafting of publications.

A special aspect of the research is the involvement of staff with varying levels of experience, often under training (undergraduates, graduate students) that may cause a high turnover. The library of documents has undeniable advantages in limiting the impact due to staff turnover, which results in the loss of personal and corporate knowledge and competencies. By formalising key processes coupled with the expertise of established researchers, a new employee may, ideally, start to contribute exactly where its predecessor left off.

An essential and proper documentation can also help in the management and storage of samples, through the registration and description of both the experiments and results, and of intermediate products, such as constructs and cell lysates, in fact practically anything filling refrigerators and freezers; all too often samples are preserved without a comprehensive description, undermining their value. It is especially important to preserve the history of the project: reports on protocols, on measures, on conditions and results, etc. Particular attention should be devoted to small, but significant, changes in the protocols, introduced in the common practice but often not reported in the documentation. When records are guaranteed by a quality system, they will be a valuable reference for any future research and a source of valid correct and complete information, should it become necessary to use the sample as well as to respond to a dispute on published or patent-pending results. On the other hand, preserving samples for years (often at extremely low temperatures)—in case it may become an important resource—must also be assessed in terms of its economic costs. The need to store samples for years can be reduced through a correct assessment of their importance; this can free up space and laboratory equipment that then become valuable resources for current projects. Business economics teaches that warehouse and inventory costs heavily affect the economic management of any institution. In any case, the management of a large number of samples is not generally the scope of a single research group, but should be managed within the context of a centralised policy—for example, the establishment of tissue banks and maintenance of animal models for the study of human diseases.

4.3.1.4 Training

Quality principles devote great attention to personnel competence and skill. In a quality system the education and training of personnel is one of the requirements to enable the organisation to achieve effectively and efficiently the strategic goals. In a starting research project, it is important to ensure that all staff have the required competences and skills to interact with the team and to perform the research work.

Very often, the research laboratory is a place of training for college students, recently graduates or doctoral students. In any case, any young researcher or any new staff member in a laboratory or in a project has to understand and acquire the standard operational procedures specific to the project's organisation. As mentioned above, having written experimental procedures allows new staff to readily align with the laboratory common references. The researcher must be able to work confidently, knowing not only the experimental procedures, but also the related controls and, more generally, the rules of laboratory management: in other words, it is important to give new researchers both scientific and quality management training.

This is a very sensitive issue, because educational institutions rarely provide students with training on management and quality. R. Davies made the point very clearly when speaking about how quality can help to improve science/scientific reproducibility. She argues that *as a result, many trainees remain unaware of how QA* [quality assurance, editor's note] *practice could elevate the profile of their work, document proof of their commitment to quality, demonstrate data integrity and illustrate how the effects of unanticipated events have been reduced or accounted for. More importantly, they lack access to effective procedures and processes that facilitate data reconstruction throughout their research program* [4].

The documents of the quality system and more specifically the documents that indicate the strategic approach to management are good training material to transfer the working method to the newly recruited. This approach prevents that the training of personnel—which in the case of research is of a high intellectual capability—be limited to the acquisition of the technicalities of laboratory methods and the following of instructions. Qualified staff can—and should—also contribute to the improvement of management and laboratory procedures, thus leaving the mark of their experience and expertise, which is of undisputable value for the laboratory or research institute.

Committing to Research Training That Covers Sound Science and Quality Practices (Rebecca Davies[3]). The reasons for irreproducible research are complex; we study complicated systems that likely contain significant variability within—and between—subjects, specimens or populations. These challenges present reasonable barriers to consistent research replication. In this exploratory environment, multiple variables may influence research outcomes in ways that we cannot control through sound science or address using best practices. However, quality assurance best practices that facilitate sound research and data management, good documentation practices and a systematic (continuous improvement) approach to research quality should be incorporated within our basic research settings and training programmes so that the scientists who work there can readily demonstrate the quality of their work and mitigate the effects of variables that can, and should, be controlled.

Most academic basic research environments and training programmes encourage scientists to develop their skills in the fundamental aspects of sound science. They focus on research planning, appropriate study design, statistical analysis, understanding bias and variables, and the responsible conduct of research (which addresses critical issues such as human subject protection, fraud, fabrication and plagiarism). This emphasis on the scientific fundamentals of hypothesis testing and ethical conduct helps scientists determine *what to do*. However, historically, we have relied heavily (nearly completely) on individual mentors (and their mentor-defined research environment) to train scientists in the practical aspects of *how to do* good science. Many mentors have significantly advanced science and successfully delivered a new generation of competent scientists. However, individual mentors are highly variable, which means that our research training programmes are highly variable as well. Since mentors drive research practices, establish research cultures and train the next generation, a high level of variability may have significant real-time and future consequences.

A commitment to research quality assurance programmes within basic research settings provides an opportunity to encourage best practices and drive research quality consistency. In addition, quality management systems are designed to ensure that records are complete, accurate, reliable and secure. As a result, research and data reconstruction are possible, providing mentors and mentees with credible evidence of the quality and reliability of their work.

The nearly universal lack of training in research quality assurance best practices constitutes a gap in our research enterprise. Because research informs future (and

---

[3] Rebecca L. Hegstad-Davies, Associate Professor, Director, Quality Central Department of Veterinary Population Medicine, University of Minnesota College of Veterinary Medicine.

helps us evaluate past) research, the entire research community remains vulnerable to the impact and potential misdirection of research data, inference and outcomes generated by those unable to demonstrate a commitment to both the fundamentals and the best practices required to conduct sound science.

4.3.1.5 Communication

The importance of effective communication within a research group goes far beyond the simple exchange of information and data. *It seems almost absurd that how we communicate could be so much more important to success than what we communicate*, writes Alex 'Sandy' Pentland in a 2012 article [5] underlining the importance of communication in a team for achieving good results. According to Pentland and the research he performed on different teams, there are three main unconventional indicators of team performance, and all three are linked to the way people interact outside formal meetings, i.e. communicate naturally. But he goes far beyond, saying *the best way to build a great team is not to select individuals for their smarts or accomplishments but to learn how they communicate and to shape and guide the team so that it follows successful communication patterns.*

Even from our direct experience we can tell how much easier it is to work with people sharing information and knowledge, through effective communication. Pentland identifies several communication characteristics that may help in observing, diagnosing and addressing the way researchers in a group interact, reinforcing their ability to achieve results together:

1. Everyone on the team talks and listens in roughly equal measure, keeping contributions short and sweet.
2. Members face one another, and their conversations and gestures are energetic.
3. Members connect directly with one another—not just with the team leader.
4. Members carry on back-channel or side conversations within the team.
5. Members periodically break, go exploring outside the team and bring information back.

The behaviour of team members affects three aspects of communication: energy (the number of interaction signs, such as verbal answers or nod), engagement (how energy, i.e. interactions, is equally distributed among team members) and exploration (communication with external subjects). Teams that have great energy and equally distributed exchanges, and seek outside connections, perform significantly better than others. Pentland gives some clues: *the most valuable form of communication is face to face. The next most valuable is by phone or videoconference, but with a caveat: those technologies become less effective as more people participate in the call or conference. The least valuable forms of communication are e-mail and texting.* Be then aware that *energy is a finite resource. The more energy people devote to their own team (engagement), the less they have to use outside their team (exploration), and vice versa.*

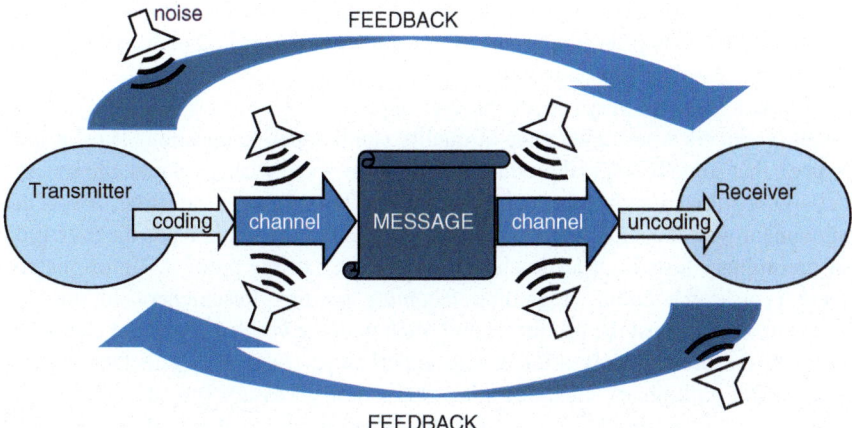

Fig. 4.5 Transactional model of communication. Information is coded by the transmitter and sent via a communication channel. The receiver can use a different channel to retrieve and decode the message, and *understand* the information. The receiver may give feedback to the transmitter. The whole transmission is affected by *noise* of various kinds

On the other hand, communication to be effective needs adequate skills such as knowledge of a communication model[4] (see Fig. 4.5), awareness of issues that can arise and ability to correctly address the transmission of information (effectiveness) spending as few resources as possible (efficacy). In the model of communication illustrated, information is coded by the transmitter and then sent via a communication channel that could be different from the one used by the receiver to retrieve and decode the message, and *understand* the information. For example, in a face-to-face conversation, the transmitter's channel is speech and the receiver's channel is hearing. The receiver may give feedback to the transmitter, not just with speech but also with smiles or other facial expressions. The whole transmission is affected by *noise* of different kinds of which environmental, emotional and cultural are but a few examples.

Communication skills and management within the research group are extremely important throughout the project, particularly in the final phase, when interacting with external actors from different fields: e.g. applied or industrial research institutions. To enhance this latter type of communication, researchers should familiarise themselves with the language used in industrial development. This is why we encourage researchers to explore, experiment and use some tools, vocabulary and concepts familiar to the final recipients of the research results. Some tools of this nature will be illustrated in Chap. 4.

4.3.1.6 Procurement

Quality assurance of the final product is strongly affected by the control of raw materials or—more broadly—of the physical resources used to make it. Many kinds

---

[4]Widely known model by Belch&Belch.

of biological and chemical products are used in experimental activities: if not adequate, they can compromise the scientific results and—as stated in different studies (see Chap. 1)—their reproducibility.

Regrettably, some very specific biological products may be provided only by few small producers whose guarantee of quality can be open to question. Thus, it may happen that some biological products do not show, even when declared, the specific activity required for the planned experiments. This would need either a specific agreement with the supplier, which in most cases is very unlikely, or the execution of preliminary tests in order to check the adequacy of the product. Unfortunately, there is a scarce culture regarding *incoming quality*. Researchers, in the first instance, tend to trust the producers and often plan the verification of the characteristics of the product only after having experienced frustration with their experiments. The preliminary check on materials is not necessary for all of them. The most critical ones should be identified in the study plan and submitted to simple experiments to ensure their suitability before starting the research experiments.

Basic research employs very specific materials (reagents, consumables) in small quantities. The researchers often face difficulties in finding the materials, and it can happen that they are forced to buy batches larger than they really need. Suppliers are usually very large and geared to partnerships with larger interests and volumes. It is therefore difficult to establish a privileged relationship with them, as demanded by the eighth principle of TQM. In addition, the difficulties of material acquisition may induce the researcher to use similar materials more readily available at the time. On the other hand, applied research, acting as the bridge from the ideas born from basic research to the industrial replies, uses materials and tools closer to those used for production, which are provided by the same supplier and through the same channels of the production world. In addition to these intrinsic issues, the reorganisation of suppliers—dictated by economics and rationalisation of distribution—is gradually erasing the peripheral warehouses in favour of centralising stocks, forcing on customers the need for an earlier purchasing programme.

Thus in basic research, we have reduced quantities, crucial product quality, critical logistics and timing to maintain the development schedule, and even economic consequences should the final product reach the market late. We can therefore understand how critical it is to manage the suppliers according to quality standards.

When laboratories are part of a larger institution, the difficulties in the supply of small lots can be solved with the centralisation of purchases. Purchasing larger lots allows for the establishment of a closer relationship with the supplier, with clear economies of scale. This matter is outside the aim of this text, but it is worth noting that a list of suppliers approved on the basis of quality, economy and service gives significant advantages to the economic management of the research project.

However, when the research laboratory is a small unit, a deliberate evaluation of supplies, qualification and planning on the basis of quality criteria is even more important, to achieve and maintain adequate provisions to the research studies. Failure to do so exposes the small laboratory to the risk of a shortage/lack of supplies, inadequate use or excessive expenditure on materials.

The control of supplies involves a two-pronged approach: evaluation and qualification of suppliers authorised to provide materials/services, and control over the goods or services purchased. In the former case, we define the criteria (usually quality, cost and level of service) used in the first instance to include a vendor on the list of *qualified* suppliers and in the second instance to periodically evaluate its performance. If a quality management system is in place, the results of these assessments are submitted to senior management during the annual review. The matter of incoming materials will be dealt with in Sect. 4.5.4.4.

4.3.2 Management of the Research Project

When starting a research from an initial idea, the researcher is creating a project, which should have several phases: a planning, an execution and a conclusion phase with the exploitation of results (see Fig. 4.6).

In the following paragraphs, we will discuss in further detail from an organisation and management perspective the three main phases: planning, execution and exploitation. The careful reader has probably noticed that this scheme recalls the classic PDCA, where controls of the check phase are included in the execution phase.

4.3.2.1 Planning

There is a common perception among scientists that planning a research project is either impossible, because of the uncertainty of the outcome, or counterproductive, because of constraints it might impose on the development of the study. One of the best answers to this reasoning is found in the QPBR, as shown in Box 4.2.

When speaking about planning, we think immediately about tasks, milestones, deliverables and all the concepts that go under the title of *Project Management*.

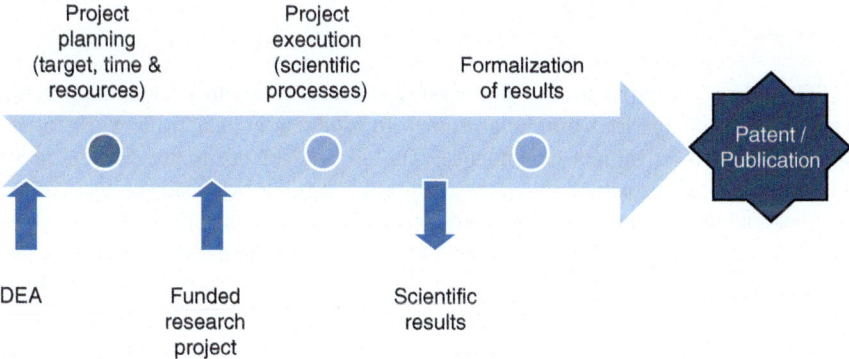

Fig. 4.6 Phases of a scientific research project management. Starting from the idea, the project is planned with respect to targets, time and resources; then once the funds have been guaranteed, the project is executed and provides scientific results. The formalisation of results leads to either a publication or a patent, or both

However, there is also a need for planning the scientific side of the study, i.e. the experiments. We will now consider both these aspects.

Planning the Study. Most funding bodies request a project plan, so researchers are familiar with the basics of Gantt charts, even though this representation is, most often, presented to satisfy a specific request rather than to effectively plan the work. The first task of a research project should be to define objectives, main activities together with related accountabilities and responsibilities, and needs for resources. In a quality-regulated environment, *to plan* means not only this, but also to foresee checkpoints. Very often checks, both on the progress of the plan and on the outcome of scheduled tasks, are overlooked. It may also be useful to identify the dependencies between activities, and which tasks are bound by previous activities or outputs—documents, materials, information or equipment. It is essential that planning becomes a tool for monitoring the evolution of the project and, for this reason, it must be kept up to date. It is also true that the planning of a scientific project requires more than any others to be constantly updated as the result of each phase is difficult to predict.

Box 4.2: About Planning the Scientific Research

The fact that responsible scientists define the processes in advance adds credibility to the whole of the study and greater acceptability to the results. In no way does this structured approach restrict the innovation or technical freedom of the scientist, nor does it prevent a scientist from modifying the plans as the study progresses. On the contrary, the structured approach helps the scientist to eliminate, control or understand the variables that inevitably impact on studies, enabling a valid and confident interpretation of the study data. In this way, sound study management contributes to the credibility of results and their scientific meaning.

WHO-TDR Handbook of QPBR—5 Conclusions

The use of a simple tool of temporal and causal relationship can support the researcher in charge of a unit or a project, to establish a clear framework for the activities and their mutual dependence. The advantages are many. Collaboration between different groups requires the temporal coordination of sub-projects aimed at transferring knowledge or intermediate products (e.g. vectors, cell lines, engineered mice) to other partners to complete the project. The management and planning of material supplies are crucial, especially when projects involve multiple research units and multiple locations. For instance, the lack of a reagent, whose availability has not been verified in due time or which has been deemed at the last moment unsuitable for the intended use, can delay the research and have an impact upon the work of more than one group or branch of activity. There is a substantial advantage in using a planning control panel: the right direction can be maintained without meandering in diverging streams, using and focusing resources and heading on most promising paths. This also allows to enlarge the field of research and to add innovative goals whenever new opportunities arise while maintaining consistency with the project and even opening new research lines well matched to the available resources. Everything is facilitated, if an organised and coherent overview of the project is maintained, and if the intermediate results are properly recorded and evaluated.

> **Box 4.3: An Appropriate Aphorism**
> If you fail to plan, you are planning to fail.
>   Benjamin Franklin

To assess the ongoing progress of the project, it is also useful to produce a grid of forecasted milestones and deliverables, a measurable and tangible outcome of the project most often required by funders. In short, the use of simple diagrams can be of great help to anticipate problems, to facilitate communication, but also for freeing up mental resources otherwise engaged in an always challenging control, to the benefit of scientific creativity and intuition.

Figure 4.7 shows, as an example, the plan for a project involving three research units. Black arrows have been added to a simple Gantt, to indicate the time dependencies of some phases from the results of others. A diagram of this kind can be represented with a more sophisticated tool such as PERT (see Sect. 6.1.5. **Network**

1	Activity on cells and tissues	units 1, 2, 3
1.1	intracellular pathways for the substance production	unit 1
1.2	effects of the substance on specific cell populations	units 1 e 3
1.2.1	effects of exposure to the substance	unit 1
1.2.2	toxic properties of the substance of various cell types	units 1 e 3
1.2.2.1	cells	"
1.2.2.2	tissues	"
1.3	effects of the substance on synaptic plasticity	units 1, 2, 3
1.3.1	changes in the signal transduction	unit 1
1.3.1.1	evaluation in physiological context (tissues)	unit 1
1.3.1.2	assessment of electrophysiological parameters	unit 2
1.3.2	influence on synaptic plasticity mechanisms	unit 2
1.3.3	treatment in specific brain areas	unit 3
1.4	assessment of activation of signal transduction pathways	units 1, 2
1.4.1	ERK activation in the identified areas	unit 2
1.4.2	PKC activation in the identified areas	units 2,1
1.4.3	state of synaptic plasticity in tissue slices	unit 2

2	Activity on animals	units 1, 2, 3
2.1	assessment of the effects on physiological functions	"
2.2	administration of the antagonist substance inhibitors	"
2.3	inhibition effect on behavior of the antagonist substance	"

Fig. 4.7 Gantt of a research project (simplified). The project is split into two main activities, on cells/tissues and on animals, and each into three more levels of detail (not all shown). Phases 1.3.3 depends on the availability of the results of 1.2, and phase 1.4 provides results for 2.1

Diagram/PERT). The breakdown of tasks and mutual relationships between units facilitates the coordination and shows the sequence of expected results.

Two different controls have to be added to a well-conceived plan: one on the intermediate results and the other on the overall timing. The first kind of controls is aimed at verifying if the project is producing the expected results, so it would consist of comparisons, verifications and tests. Design review is an example of intermediate check that has to be introduced in a plan. During the development of the project or even at the end of specific phases, the research group should make a review, possibly involving colleagues or external experts: to check if conduct and results are correct and complete; to control data gathering and analysis and to find ideas for new investigations or experiments. If a review is done at the end of a project phase, it ensures that the activities that follow will not overlook inputs and will stand on solid intermediate results. The research group should also be uniformly informed about the progress of the project (see Chap. 4—Communication Sect. 4.3.1.5).

The second kind of controls is simply the formal verification that foreseen activities have been completed in due time or that the percentage of completion is as forecast. Knowing in advance that we are going to miss or delay some objective gives the opportunity to recover, before it is too late. This might well sound as absolute common sense, but it is actually rare to find it applied to project planning.

Planning the Experiments. The RQA Guidelines on Quality in Non-Regulated Research [6] provide a summary on the preparation of a good plan for the conduct of experiments within the framework of a study; let us quote here an extract from the RQA guidelines, enriched with some new hints:

1. Define and write down clear aims and objectives of the study.
2. Identify the study, every experiment in the study and every material or animal in each experiment.
3. Randomise the subjects of the experiment groups.
4. Establish blind procedures and analysis, to avoid influencing results.
5. Take care of main statistical directions and practices: for example, dimension appropriately the size of samples, according to statistical principles; define criteria for the exclusion of outliers before analysing the results.
6. Define test criteria before running the experiments.
7. Refer to predefined rules for stopping data collection.
8. Fix intermediate checks and end points before starting the study.

These directions are not intended to impair the research process; rather, they provide it with a clear pathway of rules, checks and dimensions aimed to avoid misinterpretations, opportunity for error and ultimately reworking of experiments, as well as precise experiment designs. Among the advantages ascribable to this approach, there is the opportunity for dimensioning more precisely the experiments in terms of resources (materials, manpower and equipment).

Talking about designing experiments, we cannot fail to cite the design of experiment (DoE), also known as experimental design. It is a statistical method for planning and analysing experiments, in order to save resources and obtain the maximum amount of information. Developed by Sir Ronald Fisher in the 1930s for the optimisation of crops, it has been exploited in the (mainly chemical) industry, before returning to the biological field. See Chap. 5 for a further discussion.

4.3.2.2 Executing and Measuring

As indicated for example, in section 8.5 of ISO 9001, during the execution of a study, only validated protocols should be used to guarantee correct and repeatable results. As *validated protocols* we intend any written experimental procedure, whose results, checked in different conditions for an appropriate number of times, are within the accepted tolerance.[5]

As stated by the quality gurus Juran and Deming, the result of the main process is not guaranteed if the support processes are not controlled. Our experimental procedures use materials and equipment, and are carried out by personnel; that is why we need materials to be correctly procured and controlled, i.e. adequate, not expired, correctly stocked and treated. We need that equipment is maintained and periodically checked, and that researchers and operators have the right skills and confidence to use them. The environment also has to be suitable (temperature, humidity, cleanliness, sterility) for the purposes of the project.

It would be advisable, also, to monitor the experimental process and promptly store the most significant data to be able to control and—if necessary—to correct it, and to have available adequate storage units for the experimental outcomes. In both cases, there is the need for procedures to identify data and outcomes, so as to be able to trace back their origins and related experimental conditions. The discussion on staff skills, materials and reagents, equipment, samples and raw data will follow in the next chapter.

Finally, the release of experimental products or results should be authorised after a defined control, which could be either a dedicated design review or a set of measurements, or, better still, both. A paragraph below is dedicated to how, when and why to perform research reviews.

4.3.2.3 Exploiting

The exploitation of a research project runs along three axes: scientific publication, patenting and technological transfer to manufacturing. The benefits in structuring and performing a research study according to quality principles are evident when dealing with patentable subjects: data are back traceable, procedures are clearly documented and results are rigorously analysed and thus proven. It can be said that this is the only way of proceeding when working for a patent application. The same advantages are found when the final result of a study is a publication. In the introduction, mention has been made of serious problems with papers that had to be

---

[5] See Sect. 4.5.5.1.

withdrawn for unreliable or not reproducible results, ending in serious damage to the researcher's reputation. Having organised the study and collected, stored and reported data and results in an unimpeachable manner not only prevents scientific publications to be challenged, but also has another indisputable and positive effect: further studies based on these data and results will not waste resources by following incomplete or ill-defined directions.

Scientific results can also end in a technological transfer. The results of the basic research projects cannot reasonably be directly industrialised, because researchers, in the execution of experiments and the definition of methods, often use materials and resources only available in laboratory conditions. The inherent danger of improper material manipulation by operators, the need for many manual operations or the continuous calibrations are not constraints in a research laboratory, since all operators are extensively trained and skilled, and there are no time-related costs. Industrialising a project means to identify materials, technologies and production processes suitable for the mass production of a repeatable, reliable and cheap product. Achieving this result requires—before starting development—careful analysis of the needs of the manufacturing process, as well as maintaining results consistent with the initial requirements throughout the development of the study. In other words, if the researchers see potential exploitation of their research project, they should identify and acknowledge the constraints of industrial research in developing the product, together with the principles of quality conduct in their research. With the application of these principles, the *research product* delivered to applied research will have a higher value since a better definition of materials, processes and validations will lay down the basis for adequate quality controls and will speed up production, with clear advantages for all parties involved, including public health.

4.3.2.4 The Importance of Planned Controls (Shannon Eaker[6])

Advancing a technology from basic research within standard laboratory conditions to a commercially viable product and manufacturing process is critical to ensuring that the technology reaches patients broadly. This evaluation should take place as early as possible, and should involve many stakeholders including scientists, clinicians, operations managers, regulators and even payers. Decisions within cell therapy and regenerative medicine processes are largely derived from the biologics and bioprocess industry, harnessing a wealth of experience, and existing infrastructure. A challenge here is that many processes are standardised for a specific product, such as monoclonal antibody production. Within cell therapy, specifically autologous products, the incoming material is patient specific; thus, patient-to-patient variability limits the ability to standardise processes across the board. However, best practices are needed to ensure that processes are at current good manufacturing practice (cGMP) standards (such practices include technology transfer, cell sources, isolation and cryopreservation procedures, characterisation/phenotypic assays, quality assurance among others).

---

[6] Cell Therapy Enterprise Technical Leader, GE Healthcare, Knoxville, Tennessee, USA.

Many important steps and critical pitfalls in taking a research project to market can be unveiled early in the process. Discussions with institutional business development and technology transfer stakeholders often yield undiscovered opportunities, such as partnerships or research collaborations that can be developed. Interactions with regulators early in the process are often desirable, and could shorten the commercialisation time frame by avoiding unnecessary risks. Early interactions allow the regulators to become familiar with the technology, and can identify potential risk areas. This is also an ideal time to determine if there are commercialised products already under evaluation, learning from other projects that might be similar to the project being evaluated. Even if a specific process is vastly different from standard process, desirable overall features still remain. Scalability of the process, specifically within equipment and platforms, is important in process evaluation. Determining how many patients to be treated once fully commercialised will determine the scale of the manufacturing process. Determining the input and output cell characterisation assays is also task to be evaluated. As the research project moves to manufacturing, what potential steps could alter the cell phenotype?

It is important to evaluate areas where variation might occur, specifically around cell potency and phenotypic. For example, if specific cell surface markers are required in the final product, monitoring these values throughout the process is desirable. A specific procedure might alter the phenotype, and must be addressed early in the process. As another example, if potency testing uses assays that monitor cell killing, determining specific steps that alter this assay must also be determined. A balance with maintaining the desired cell product with manufacturing optimisation must be addressed and followed throughout the entire process.

Properly documenting and reporting negative results is required, and often leads to more optimised manufacturing strategies including predictive analyses.

To summarise, *determining the factors that can enhance the success of a cell therapy product early in development is critical to the overall success of the product. Even if all of the processes described above are not achievable, addressing them in the development stage will benefit downstream interested parties. Making data available, whether positive or negative, is a critical component to success* [7].

4.4 International References and Models

As illustrated in Chap. 2, there are no official international references for quality in basic biomedical research, as found in the GxPs, which rule the subsequent phases for the development of drugs or therapeutic protocols. In addition to the generic international standards outlined in the paragraph dedicated to the quality standards, the researcher can find excellent guidance in some documents issued by international bodies, universities and councils. Many examples can be found in large number in the UK, and few in other countries (e.g. Spain); most of them are called good research practice (GRP). There are examples of funding bodies that request the application of a quality code to submissions, and universities and research institutes that develop their own guidelines. There are examples that follow the structure of a standard quality system; most frequently, they are divided into chapters covering the main operational aspects, at times extending into subjects such as publication management and ethics.

4.4.1   WHO-TDR Handbook on Quality Practices in Basic Biomedical Research

The most complete and wide ranging is the WHO-TDR Handbook on Quality Practices in Basic Biomedical Research (QPBR), known and applied worldwide. The QPBR was first issued in 2001 for the management of research in tropical diseases (TDR Special Programme for Research and Training in Tropical Disease), and was revised in 2006. The handbook was intended to bring scientific research carried out in those countries where tropical diseases are endemic, to the same standard as that of the most advanced countries; despite this specific origin, it is appreciated and considered a reference even in industrialised countries.

Box 4.4

The quality practices for basic biomedical research described in this document do not address the scientific content of a research programme or proposal, but are concerned with the way the research work is managed.
   WHO-TDR HANDBOOK ON QPBR, Foreword
   Meaningful scientific interpretation of study results is only possible when founded upon reliable data. Clearly, in order to obtain reliable data, the experimental variables that always affect studies must be kept under control. Quality practices are designed to help scientists control the variables. Such an approach is the only way to obtain reliable results and sound scientific interpretation, and to avoid being dogged by 'false positives' and 'false negatives'. This reasoning explains why best practices attach so much importance to precise planning (through the written study plan or protocol) and to use of standardised techniques (by following previously written standard operating procedures).
   WHO-TDR HANDBOOK ON QPBR, Chap. 3.1

The principle underlying the quality approach to research of the QPBR is shown in Box 4.4. As stated by the authors, the quality approach in basic biomedical research refers only to the management, deferring the assessment of the value of the research itself to a different context. On the other hand, a well-structured approach to research is expected to ensure that results comply with the basic principles of the scientific method, namely reliability, reproducibility and traceability.

The QPBR, after an interesting introduction to quality in research, is structured in chapters that follow the main topics of research as listed below:

- Organisation
- Physical resources
- Documentation
- Supervision/quality assurance
- Publishing practices
- Ethical considerations

Finally, the appendix brings to the reader a set of documentary examples, to help in understanding and implementing, wholly or in part, a research management system. In detail, it provides templates for standard operating procedures (SOPs), curricula vitae and training records, and some sample SOPs.

4.4.2 Joint Code of Practice for Research: JCoPR (UK)

The Joint Code of Practice for Research [8] is a mandatory reference for most public funding bodies in the UK.[7] First released in 2003, it was revised following the results of a 2-year survey which started in 2006; it was then reissued in March 2012. It is based on the principles that *without QP [Quality of the research Process—editor's note], confidence in the research findings is much reduced* as is *fitness for purpose*, and it fosters quality of the research process and quality of the science.

Contractors to the Department of Environment, Food and Rural Affairs (DEFRA) as well as other public funding bodies have to accept the terms of the Code and demonstrate their application by means of internal and third-party audits and, on occasions, of the funding agency's audit. It should also be reminded that contractors have the opportunity to discuss with the funding body any item they consider inappropriate or not applicable to their field.

Within the JCoPR, main areas of quality process management are:

- Responsibilities of the contractor organisation and their managers, group leaders, supervisors and principal investigator
- Competence of the researchers involved (e.g. training, records)
- Project planning: which must be in writing; up to date; regularly reviewed; integrated with ethics, methods and resources; and agreed with the funding body
- Quality controls: both on scientific results and on processes and procedures
- Health and safety (compliance with regulatory requirements)
- Handling of samples and materials: defined in the project plan, properly identified, stored and disposed
- Facilities and equipment: appropriate, safe, identified, maintained, calibrated and provided with SOP for correct use
- Documentation of organisational procedures and scientific methods, the latter properly validated

---

[7]The code is requested to be applied by contractors of DEFRA, the FSA, NE, VMD, AHVLA, MMO, Fera, FC, the UK Devolved Administrations, BBSRC and NERC.

- Research/work records, detailed enough to allow repeatability; regularly reviewed by the principal investigator (PI); properly stored and protected against unauthorised modification
- Field-based research, compliant with environment legislation

Again, the sole interest is on the research process, the areas of organisational processes and governance of the research institution being outside the scope of this quality management process. However, this is a great example of a public institution playing the proper governing role in educating and spreading the culture of quality management in the field of scientific research.

4.4.3   Research Quality Association Handbook and Guidelines

Research Quality Association (RQA) was founded in 1977, with the purpose *to drive quality and integrity in scientific research and development* in the fields of pharmaceuticals, agrochemicals, chemicals and medical devices. As stated on the RQA website,[8] *since its beginnings in 1977, the Association has grown and developed to reflect regulatory changes, the impact of regulatory inspection and the changing structure and needs of industry.* In recent years, the RQA has begun to address quality in the non-regulated research, issuing a handbook for quality systems and a booklet with a summary of the quality requirements applicable to a research laboratory. While the *Quality Systems Workbook* (issued 2013 [9]) is freely available on the RQA website, *Quality in Research—Guidelines for working in non-regulated research* (last issued Sept 2014 [6]) are available for purchase at a low price.

The Quality System Workbook guides through the main requirements of the ISO 9001 standard and provides useful examples and exercises to better understand and apply them. If quality references are sought without the need to adhere to international standards, the Guidelines represent a concise compendium that in 11 short chapters outlines the quality requirements to establish a controlled lab and to conduct quality-regulated research, together with very useful hints on related risk and opportunities. In Chap. 3 such requirements are widely integrated with directions from other sources.

4.4.4   Other Examples

The UK Research Integrity Office, UKRIO, is a small, independent charity, staffed almost entirely by volunteers. It offers *support to the public, researchers and organisations to further good practice in academic, scientific and medical research*, promoting *integrity and high ethical standards in research, as well as robust and fair methods to address poor practice and misconduct.* UKRIO provides both dedicated

---

[8] www.therqa.com.

advisory services and free publications. Among the latter, there is a Code of Practice for Research and a condensed, yet useful, non-technical checklist for researchers. The Code has been adopted by many research organisations, including over 50 universities. Though UKRIO is active only in the UK, its website and publications can be a reference for researchers all over the world.

Research Council UK (RCUK) is a coordinating body between councils, which aims *to optimise the ways that research councils work together to deliver their goals, to enhance the overall performance and impact of UK research, training and knowledge transfer and to be recognised by academia, business and government for excellence in research sponsorship.* RCUK Policy and the Code of Conduct on the Governance of Good Research Conduct [10] have been submitted to the attention of all heads of universities, colleges, research council institutes and RCUK-recognised research organisations.

Other significant issues of good research practice are found in the UK, some of them devoted mainly to research integrity and ethics:

- University of Cambridge, Good Research Practice: http://www.admin.cam. ac.uk/offices/research/documents/research/Good_Research_Practice.pdf
- University College London (UCL), Code of Conduct for Research (referring to RCUK documents): http://www.ucl.ac.uk/srs/governance-and-committees/res-gov/code-of-conduct-research
- University of Reading, Code of Good Practice in Research: http://www.rdg. ac.uk/UnivRead/vb/RES/qar_public/index.htm
- Wellcome Trust (2005), Guidelines on Good Research Practice: http://www. wellcome.ac.uk/About-us/Policy/Policy-and-position-statements/WTD002753. htm
- UK Medical Research Council (2005), Ethics Series: Guidelines on Good Research Practice: http://www.mrc.ac.uk/news-events/publications/ good-research-practice-principles-and-guidelines/
- Health Research Board, HRB Guidelines for Host Institutions on Good Research Practice: http://www.hrb.ie/research-strategy-funding/policies-and-guidelines/ guidelines/guidelines-on-good-research-practice/

These examples illustrate that in the UK the culture of good research management and conduct is widespread. Outside the UK, we find an excellent example in Spain. Both the University of Barcelona and the Universitat Autònoma de Barcelona issued their codes of good research practice, respectively, *Codi de Bones Pràctiques en Recerca* or *Code of Good Research Practice* issued in 2010, and the *Code of Good Practice in Research*, in 2013. In the Universitat de Barcelona, the enforcement of the Code is promoted and supported by an internal Research Quality Office, the *Servei de Qualitat de la Recerca*.

Last but certainly not least, the *European Code of Conduct for Research Integrity* [11] was issued by the European Science Foundation. Once again, the aim is ethics, focused mainly on research integrity, but on close examination it also gives good directions and advice to the conscientious researcher.

The *Servei de Qualitat de la Recerca* (Carmen Navarro[9]): The Universitat de Barcelona (UB) has a long experience in the implementation of quality management systems applied to research. Since 1994, first with the quality assurance (QAU) and afterwards with the *Servei de Qualitat de la Recerca* (quality research service, QRS), the research groups and the core facilities at the UB receive support and advice for all the processes of implementation, formal acknowledgement and maintenance of quality systems. As a result, ISO 9001, ISO 17025 and good laboratory practice certifications/accreditations have been achieved.

The support given by the QRS staff to the researchers includes the following:

- Advice in selecting the quality system
- Support in the process of implementation
- Conduction of audits/inspections
- Support to external inspections/audits
- Training of laboratory personnel in quality assurance systems
- Formal review and controlled distribution of standard operating procedures (SOPs)
- Custody of certified standards for apparatus calibration

One of the main concerns of the QRS is the promotion of quality in all the research activities at the UB in order to assure that the results obtained are accurate, reproducible and traceable. For this reason, in 2010 the QRS promoted the creation of the Code of Good Research Practices [12], which establishes the guidelines for carrying out research activities of any kind at the UB. Its objectives are to improve the quality of research in all fields; to set up mechanisms that ensure honesty, responsibility and rigor in research; and to ensure that junior researchers acquire good scientific practices.

The document, although initially designed for assuring the quality of the results of research, in its final version also includes some guidelines regarding integrity in research, as quality and integrity in research are actually two sides of the same coin.

The organisational structure established at the Universitat de Barcelona, with the Quality Research Service, allows compliance with different quality standards, and provides the tools for improving the reliability of scientific results.

4.5 Quality Management in Basic Biomedical Research

In this section, we are going to consider various aspects of research management, focusing on its application in a generic basic biomedical research unit, institute or laboratory. We will take a glance at the organisation and focus on how to manage the process of research and the support processes, namely management of staff, equipment, facilities, materials and documentation. We will follow the directions of the main texts on quality in research outlined in the previous section.

---

[9] Head of the *Servei de Qualitat de la Recerca*, Quality Assurance Agency, Universitat de Barcelona.

4.5.1 Organisation

An organisation is a group of people collaborating to meet a need or to pursue collective goals. To do this, it has to be structured and managed. The first concept is about defining activities and relationships, and related roles, responsibilities and authority. The second concept, i.e. management of an organisation, is about defining strategic directions and goals, providing resources and setting standards.

4.5.1.1 Quality Policy and Objectives

Let us start from the directions: ISO 9001 standard and, in the field of research, the QPBR suggest that the top management—in our case the research institution or the laboratory director, or the principal investigator—writes a quality policy. This document should describe management's engagement in pursuing a quality approach to studies, and should clearly indicate the quality practices that staff must follow, as well as refer to guidelines and operational procedures at all levels and for all activities. This strong stance—demanded of management—is underpinned by the second principle of TQM, which requires the leadership to have a supporting and motivating role.

The quality policy should be accompanied by a few, essential strategic objectives and related metrics, by means of which the achievements can be measured and demonstrated. For an effective definition of metrics, we refer to the paragraph dedicated to controls and indicators.[10] The document itself should be short, readable and concise. It is advisable that top management introduces it to the staff and posts it on the walls, publishes it on the website, displays it at the reception and even attaches it to general documents sent beyond the institution.

The UAB—Universitat Autònoma de Barcelona—publishes a fine example of quality policy signed by the deputy executive administrator for research [13]. It clearly states the *mission* (what the organisation is intended to do), the *vision* (the strategic goals), the funding *values* of the institution and finally the commitment of senior management to quality.

4.5.1.2 Organisational Chart and Job Description

The main activities of an organisation should be divided by area of competence and each area should be assigned a person in charge. This leads to a chart that is a visual representation of how an organisation is structured and that outlines the roles, responsibilities and relationships between the various roles within the organisation.

Examples of organisational management include the following figures:

- Dean of Faculty
- R&D Director
- Finance Director
- Research Governance Board/Committee

---

[10] See Sect. 4.3.1.2.

- Senior Management Team
- Principal Investigator

An organisational chart should be designed also for each scientific project, and integrated with the relative authority lines and communication channels, outlining the accountability and/or responsibility of each team member. All these aspects have to comply with the scientific, technical and support tasks.[11]

4.5.2   Quality System

A quality system is a management system. Among various definitions of management system, we report that by ISO (International Organization for Standardization):

> A management system describes the set of procedures an organization needs to follow in order to meet its objectives.

Actually, to be more specific, we refer to the set of roles and responsibilities, procedures and resources that enable an organisation to operate. The holistic attitude that characterises a management system *provides a disciplined approach for ensuring that the research, products and services supplied by the research institute are fully aligned with the quality and compliance standards of funding bodies and customers.* In other words, it *ensures that quality is built into the research culture* [6].

To be effective, a quality management system (QMS) must be tailored to the institution acknowledging its characteristics, its functioning, and the rules and good practices already in place; otherwise it becomes a forced superstructure that can barely be integrated in the day-to-day practices. Moreover, the QMS in a research structure—as well as in organisations of every kind—should be accurately designed, so as not to become a framework for bureaucracy, rather than a concrete help to management. This aspect is strongly felt by researchers who often perceive the quality rules as a *cage* for their creativity. The ISO 9001:2015 standard offers a good support to the fitness-for-purpose approach in designing a QMS for a research institution, i.e. the *risk-based thinking.* Performing a risk-based analysis can help in defining the correct level of quality procedures with respect to the requirements of the context and the balance of time/cost against the dimension and *invasiveness* of the procedural frames. A good starting point could be the researcher's latent requirements,[12] i.e. what can facilitate beyond the expected, the researcher's job. As examples, some of the more challenging tasks for a researcher are to deal with huge amount of data, or maintain the control/traceability of experimental procedures and results throughout the research project. A QMS that can respond by streamlining those challenges consequently has a very good chance of not only being accepted, but also being warmly welcomed.

---

[11] See Sect. 4.5.3.1 below.
[12] See Chap. 2—Total Quality Management: The Requirements.

Organisations are living entities, progressing and modifying their characteristics adapting themselves to the evolving external context. This is the reason QMS must be periodically reviewed in order to maintain its capacity to govern the organisation and be a reference for everyone acting within the organisation. These reviews can also take into account the feedback of the staff and foster the improvement and streamlining of the QMS, thus facilitating its application. The ISO 9001:2015 as well as the RQA Guidelines [6] state that a periodic review of the QMS assures that the system is well driven by the final recipients' requirements and makes improvement a constant process.

4.5.2.1 Quality Assurance and Quality Control

Sometimes quality control (QC) and quality assurance (QA) are referred to as synonymous, but this is not the case. Assurance is the act of giving confidence, while control is more about inspecting to verify compliance to references. QA consists of the planned and systematic activities to independently guarantee that the defined requirements for the process will be fulfilled. QC is a set of observation techniques and activities focused on identifying defects in the actual products or service provided. The goal of QC is to identify defects after a product is developed and before it is released. Thus, the main differences are that QA is process oriented and focuses on defect prevention, while QC is product oriented and focuses on defect identification and correction. QC may be used to improve the quality of results in an organisation, inspired by the corrective and the preventive actions; however, this approach is not advisable, because it is mainly reactive and not holistic (i.e. it does not consider all processes as a whole).

Usually QA is performed by people not involved in operational activities, i.e. not involved in production of the goods or services. A good example of the role of QA can be found in GLP. A quality assurance programme is entrusted to people not involved in the research. Its aim is to audit the proper conduct of the study, including data, materials, equipment and record management, to approve the documents of the study and to verify the final reports.

4.5.2.2 Research Reviews

Research reviews are a fundamental aspect of research quality. They are, at the very least, performed before the release of scientific product (e.g. a publication or a patent application), to ensure that results are robust. In addition, it is also advisable to undertake a research review when critical decisions have to be made based upon a specific result, which must be solid and reliable. Periodic reviews are also useful in maintaining the focus of the study in progress and assure good data processing, recording and analysing.

The RQA guidelines [6] distinguished three types of research reviews:

- Of the science: A peer review verifies the consistency of scientific hypothesis and conclusions.
- Of the data: A QC of the data assures that results are valid and reproducible and meet the project requirements.
- Of the processes: A QA audit on the research system guarantees that the QMS is well run and maintains the adequate support to research management.

Regardless of their type, these reviews should be done periodically, preferably with the participation of people other than the researchers involved in the study, and their reports should be properly recorded. Keeping review documentations can assure that data and decisions are easily tracked back.

The QPBR and most GRPs as well as JCoPR require formal and regular research reviews. Senior research management should be in charge of planning the research reviews and their signature on the documentation *should attest that sufficient reviews at all levels have been made to verify the data* [1].

4.5.3 Personnel

A clear definition of roles, required competences, communication among team members as well as a good working environment with a positive and collaborative spirit are crucial to the success of the research project. As suggested by the Code of Good Research Practices of the University of Barcelona, *research team leaders must take care of the working environment to ensure that the personnel can cultivate their knowledge and skills; knowledge exchange is also promoted, in order to share and attain common research objectives* [12].

On the other hand, the RQA guidelines [6] raise attention to the reputation of the research team, stressing that it is up to the management to maintain skills and competences monitored and updated, and to allocate the appropriate skills to the areas that need them. To do so, it is useful—thus recommended by different standards and guidelines—to maintain a record of staff skills, training and qualification.[13]

The turnover in a basic research laboratory is typically high, because of the presence of young people, with different levels of pre- and postdoc formation, for only a limited time frame. Study results, protocols and laboratory knowledge should then be considered a strong heritage of the institution and should be protected against undesirable or uncontrolled accesses. For this reason, everyone having access to the laboratory should be authorised by the person responsible for (e.g. the principal investigator), trained to ensure the correct behaviour regarding both relationships and scientific skills and conduct, and their presence documented in order to maintain the traceability. The signing of a non-disclosure document may be advisable.

4.5.3.1 Responsibility
There should be a clear definition of how responsibilities are to be shared and assigned, to avoid the risk of both grey zones, i.e. activities not clearly attributed, and multiple responsibilities, i.e. two roles assigned to similar activities. Each role defined in the organisational chart (either of the institution or the research team) should be described and recorded in appropriate documents specifying responsibilities, skills, knowledge and qualifications if necessary. This information should be reviewed and maintained up to date, at least in the following fields [6]:

---

[13] See Sect. 4.5.3.2.

- Education and certifications
- Career summary, including research and employment history, achievements and relevant past experiences
- Scientific/technical training
- Publications and presentations

4.5.3.2 Competence, Skills and Motivation

In a work of high-intellectual content such as scientific research, it is crucial to take care of the skills and abilities of the people involved in projects. In the present competitive scientific environment, credential of researchers must be prized and maintained up to date, in order to be *as robust as science itself* [6] and to ensure that *the quality of the results is not compromised by the inexperience of the researcher* [8]. The management of the research unit should compare periodically the available skills and capacities with those required by the roles in the institution and in the research projects. Whenever possible, the management should provide training to fill the gaps and fine-tune the professional profiles.

Special care must be devoted to safety training, since materials and techniques used in laboratories often require careful handling and application. Each country has its own safety code and each research institute has its own internal safety procedures. Staff in training must also be informed of the legal implications and must be supported by experienced personnel who follow them with regular meetings throughout the research.

Scientific and technical training is of unquestioned importance, but it is not the only field of expertise growth. People working in a multidisciplinary, international, multisite team, as is more and more often the case in research projects, need to have some managerial abilities. For many years, the international institution EMBO, responding to the growing needs of managerial skills, has been providing courses for laboratory management, dealing with staff selection, leadership and delegation, effective problem solving and communication.

Often researchers put passion and enthusiasm in their job, but, to ensure that their contribution to the research is complete, motivation must be fostered. We can recall the third principle of TQM, which states, *People at all levels are the essence of an organisation and their full involvement enables their abilities to be used for the organisation's benefit.* Human environment, access to equipment, funds or both bureaucracy and relationships are few examples of the non-technical aspect that—if overly demanding—can undermine energy and initiative, also endangering personal growth. Again, tackling the non-technical aspects needs managerial skills. There are different ways to enforce the involvement of people: from training to recognition of commitment and results.

In the mid-1950s, the psychologist Abraham Maslow proposed a motivational model of human development based on the *hierarchy of needs*, shown in Fig. 4.8. The fulfilment of basic needs is the condition for higher needs to become manifest. Basic needs for survival are at the base of the hierarchy, while towards the top are the intangible needs. Work is an important element in the process of self-actualisation and remuneration is perceived as a very important part of work but, unexpectedly,

Fig. 4.8 Maslow's hierarchy of needs. *Basic needs* include physiological and safety necessities. Once the necessities related to survival are satisfied, *psychological needs*, related to emotions and feelings, become manifest. On the top are the *self-fulfilment needs,* concerning the realisation of one's own potential

salary is placed at the bottom of the pyramid, among the basic needs.[14] Self-actualisation requires the fostering of motivation in the individual. In the work context, this motivation is promoted and sustained by encouraging education, providing engaging and fulfilling work as well as a congenial work environment. These aspects have been proven to have a greater effect on motivation than—almost any—monetary compensation.

4.5.4   Physical Resources

4.5.4.1 Economics

Among the documents and guidelines on quality in research we are analysing, only those produced by the UK Research Integrity Office (UKRIO[15]) deal with the economic aspects of the research in its *code of practice in research*, even though the topic is crucial for the conduct of a study. According to UKRIO, the main concern is on the ethical use of financial resources: the researcher should ensure to be

---

[14] See also [14].

[15] ukrio.org.

compliant with the requirements of funders, to have access and rights on financial resources and *to comply with organisational guidelines regarding the use and management of finances allocated to research projects.* Finally, the researcher should *cooperate and report any irregularities or concerns to the appropriate person(s) as soon as they become aware of them.*

Box 4.5: A Problem with Distilled Water Supply
A research institution experienced an abnormal failure rate in experiments with cells. Investigations showed that the problem was common to several laboratories. Chemical and biological contaminants were checked, as well as materials and reagents. Data from different experiments by the various laboratories were thoroughly analysed. The cause was finally found in the demineralised water, which was contaminated somewhere along the pipes. Separating the distilled water pipes from the demineralised water ones finally solved the problem. Detailed analysis of this example of problem solving can be found in the Sect. 5.4.4 *Example of Application—Contaminated Water in a Research Laboratory.*

4.5.4.2 Facilities

The management of the facility must be first of all aimed to protect people and to guarantee the quality of experimental results. The working environment should be effective, secure and safe [6]. A poor management of the research environment can cause mix-ups (of studies, test systems, test items, data), confusion or cross-contamination (for example, between chemical compounds or strains of microorganisms), seriously compromising the studies. Here are some aspects to be taken into account when managing the research environment:

- Allocation of space: Clearly define the space allowed for different studies and to different activities, for example, separate office, laboratory and storage.
- Tidiness and cleanliness: A clean, uncluttered and well-maintained workbench helps in minimising the possibility of errors: mix-ups, contaminations, loss of materials and samples.
- Special handlings: A particular care has to be devoted to radioactive and biohazardous materials. The QPBR dedicates an entire paragraph (4.6.1 Biosafety) to the safe management of biohazardous materials, while every research institution has its own code for handling radioactive materials.
- Environmental control system: Control of temperature and humidity can be essential to some types of bioresearch and to preserve the reliability of precision equipment. Some environmental filters may also be necessary to make clean air.
- Water and power supply have to be guaranteed in the correct quantity and characteristics (see an example of critical water supply in Box 4.5).

- Most delicate and valuable equipment should be connected to an uninterruptible power supply (UPS) to either preserve them from damages due to unexpected power blackout or guarantee the completion of the analysis of particularly valuable samples.
- Authorised access to the laboratory: The research laboratory can be subject to a high turnover[16]: students, undergraduates, graduate student, postdoc and visiting fellows can have access to confidential matter. It would be advisable to have written rules and permission documents to access the laboratory. In Box 4.6 a simple example of directions for external users is shown.

Box 4.6: Directions for the External User (Example)
- Follow the instructions of the tutors or the laboratory management and refer to them for any need or contingency.
- Comply with the obligation of confidentiality about research and production processes, products or other information during and after the access to the laboratory.
- Respect the procedures of the research institution and the rules of hygiene, safety and security.

4.5.4.3 Equipment

Procedures obtained through equipment: The reliability of the equipment guarantees not only good data, but also efficiency: sound, well-calibrated equipment minimises the risk of making poor decisions, arriving at erroneous conclusions based on inadequate information/data or having to rerun experiments because of doubts or mistakes. Overall, poorly maintained equipment opens the way to publishing unreliable data with subsequent loss of credibility. Having a good instrumentation registry, recording who, where, on what and when the equipment was used can help in tracking research information and even possible mistakes.

The equipment are chosen to answer specific, unanswered questions in science and must meet scientific requirements for accuracy, precision, robustness and measurement intervals. Every instrument should be identified with a label or plate carrying an ID code, listed in a logbook that contains, as well as the ID code, information regarding the model, supplier, technical characteristics and every other useful information for identification and correct use. If no one has been designated to be in charge of supervision of all equipment, it is advisable to assign a responsible person to each device. This person would be tasked with cleaning, recording, calibrating and managing the consumables. The equipment logbook should also contain (a) planning for preventive maintenance—such as calibration, cleaning and replacement of consumable parts—and (b) records of emergency maintenance.

---

[16] See also Sect. 4.5.3.

Box 4.7: Accuracy and Precision

Even if accuracy and precision may seem the same, in technical language they have two distinct meanings: accuracy is the ability of an instrument to provide the correct value, while the precision is the uncertainty on the value provided. An instrument can be accurate but not precise when repeated measurements of the same physical quantity are on average *centred* on the real value, but very scattered. It can be precise but not accurate, if repeated measurements of the same size are poorly dispersed, but the average is closer to the real value. Imagining throwing arrows at a target (Fig. 4.9): the throw is accurate if the arrows are widely dispersed throughout the target but equally around the centre. The throw is precise if the arrows are all nearby, though far from the centre of the target.

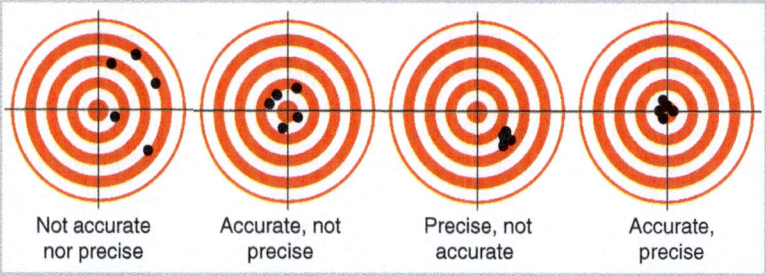

| Not accurate nor precise | Accurate, not precise | Precise, not accurate | Accurate, precise |

Fig. 4.9 Accuracy and precision. *Precise* shots are repeatable but are somewhat away from the centre of the target. *Accurate* shots lay near the centre of the target but are less repeatable

Different aspects are crucial to keep the effectiveness of the available technologies and get the most out of them. Besides the initial investment for the instrument, the host institution must consider the running costs of operation and maintenance, a substantial and long-term financial commitment. Preventive maintenance is necessary to limit the risk of finding an instrument unusable or, even worse, unreliable. Calibration of equipment used under controlled environment conditions must be regularly scheduled, usually as indicated by the manufacturer and depending on the frequency and conditions of use. When equipment is used outside the laboratory in field conditions, calibration may not withstand the changed environment; thus it should be periodically checked before being used again. In this case, the frequency of controls and calibrations must be adapted. In both cases, if accuracy and precision (see Box 4.7) are the most important features of the instrument, it is worth maintaining a record of its test measures, in order to identify negative trends that could compromise its correct use and therefore indicate that recalibration is required regardless of established schedule. Box 4.8 shows a simple example from the QPBR.

As part of an intelligent maintenance system, recording malfunctions and restorations, statistics on failures can draw attention to the need for major interventions or substitution. Qualified personnel must conduct every preventive or extraordinary maintenance to avoid manufacturer's warranties being invalidated or causing further damage to the instrument. If the equipment is valuable, unique or fragile, an unexpected failure could seriously affect the research. The JCoPR suggests preparing a contingency plan outlining how to act and where to find adequate backups, thus limiting the impact of adverse circumstances.

An instruction leaflet should accompany all simple instruments, while equipment that is more complex will need an instruction manual. Whenever instruments are available for researchers to use for their experiments, dedicated personnel must provide the necessary expertise through hands-on training and direct assistance to design experiments. In this case, the equipment logbook should contain a list of authorised personnel. Instructions should also include the rules for safe use and directions for proper cleaning (when applicable).

It may be advisable to record access to most significant equipment via a dedicated logbook. This record is useful to plan access and availability, to know when, how and how much it is used, to track possible causes of misuse and to pre-empt the supply of related consumables, if needed.

GLP brings together under the term *facility-based inspections* the periodic checks required on the equipment and on the project environment in general, to verify the level of compliance to standards.

Box 4.8: Starting by Calibrating a Balance
Calibration of balances is a good place to start (because one can follow up by calibrating volumetric equipment). Typically, the institution has a set of standard weights with which to calibrate the balances—at least once a year (secondary calibration). If there is an accredited standardisation organisation in the country, representatives check the institution's standard weights once every 3 years (primary calibration). In addition, daily checks with one standard weight will ensure validity of the weighing.
WHO-TDR Handbook on QPBR—4.2 Physical Resources

4.5.4.4 Products for the Research

Good research results are based on the appropriate use of many different products (including labware, reagents, solvents, culture media, buffers, biochemicals, kits, etc.) and their good management. Poor-quality products, their inaccurate documentation or management may seriously compromise the scientific results hindering the search for the causes of a failure. The JCoPR suggests planning the use of materials and their relative controls. Maintaining reliable information about what has been used for an experiment also enables to reproduce it allowing for selection of equivalent materials.

Box 4.9: Qualifying a Supplier

A very good example of supplier qualification is offered by the cGMP, namely the good manufacturing practices updated with the latest technologies and systems; these being the GMP require a detailed qualification process for suppliers, primarily from the active pharmaceutical ingredient (API). The qualification process may be organised in the following steps:

- Selection of supplier, according to specific criteria, for example: specification and quality, innovation, logistic (lead time to produce, delivery time, level of service, etc.), cost.
- Due diligence process, i.e. a formal comparison among several potential suppliers, with reference to the selection criteria: An on-site visit (audit) can be conducted for a deeper evaluation, as well as analysis and tests on the manufacturing line to evaluate compatibility of the material(s) with the production process.
- Assessment of all potential suppliers according to quality requirements: system/product certification and compliance, supplier audit, contract/quality agreement.
- Registration of qualified supplier(s).
- Monitoring and periodic reassessment based on defined metrics, for example: number of *out-of-specification*, on-time delivery, number of rejected batches.

Materials can be classified in categories needing different levels of rigour in assessment, for example: non-critical raw materials, critical raw materials, critical intermediate materials and API. A useful guide can be found at http://apic.cefic.org/

Some products can be purchased or produced *in-house*. In the first instance, the goods must be carefully considered before acquisition, identifying reliable suppliers and describing accurately the goods to be purchased, as well as ensuring correct and documented conditions for their transport during the delivery. Following the example of ISO 9001 standard, suppliers should be evaluated before purchasing from them and only included in a list upon a formal qualification (see Box 4.9). Suppliers in the list are then to be considered the secure choice and should periodically be reassessed to verify that they are maintaining their standards. This best practice is not easily applicable in a research laboratory: frequently, ordered quantities are very small; occasionally materials are received as small samples; often purchase is made through a dealer and not directly from the producer and it is sometimes the case that materials are exchanged between the various laboratories in a spirit of collegiality. When specific products are obtained from colleagues or unknown suppliers, a careful pre-assessment should be made before using them for experimentation. The risk of procure inadequate products and services can be reduced by maintaining a list of reliable and periodically reviewed suppliers.

Once the supplier has been chosen, it is time to order. There must be adequate documentation that includes all requirements and other useful information—for example detailed supplier's codes or a request of a complete characterisation. Upon receiving the purchased materials, the laboratory should control both administrative and technical documents to ensure that they are all compliant with the order. ISO9001 requires adequate recording of such controls.

After that, the products should be identified so as to be referred to in a secure manner. The easiest way to do it is to label them. The label will carry all information known upon arrival in the laboratory and other useful information collected during their use, for example, expiry date, opening date, required storage, limitation of use and care instruction or reference to them. The GLP requires that chemicals, reagents and solutions be labelled to indicate identity and concentration if appropriate; expiry date and specific storage instructions are also required, as well as readily available information regarding source, preparation date and stability.

GLP refers also to solutions, i.e. intermediate products generated in the laboratory to be used in experiments. The solutions prepared in the laboratory deserve the same care as the research product, because their quality can directly affect results. They have to be prepared according to established and validated methods,[17] using controlled materials and reagents. Their preparation should include directions for handling and storage as well as the expiry date. As far as expiry date is concerned, it would be more advisable to run some assays to test the material stability and define adequate storage conditions than to decide a reasonable time and storage based only on experience. Finally, if storage conditions can seriously endanger the quality of materials, all significant conditions of storage should be recorded. For example, if temperature is critical, material should be stored in a data-logged refrigerator.

The Code of Good Research Practice of the University of Barcelona suggests also to be careful with solutions or other products obtained from external laboratories, and to use them only under a signed *protocol on the corresponding transfer.*

It can be useful to keep records of the product usage for better planning of the laboratory supply and to support possible investigations of failure. This kind of record can be limited to specific and critical reagents and materials, but a good record of such information is most helpful. An adequate planning of product purchasing can prevent shortages as well as rationalise the use and efficiency of goods, which may result in overall savings; this is why—when possible—the recording of material usage should be assigned to an accountable person.

This approach becomes more significant in terms of savings, and thus very important, for large organisations and should also be expanded to equipment management. At the end of this chapter is reported an example of a centralised management of materials and instruments in a large scientific institute.

---

[17] See Sect. 4.5.5 below.

Generic and special waste management is usually defined by the regulations of each research institution and is subject to national laws. For this reason, it will not be considered.

4.5.4.5 Computer Systems

Computer systems are a special part of the equipment, used for collecting, elaborating and analysing data, writing reports and storing information, and as such deserving a special care.

One of the worst nightmares for a researcher is losing experimental data. The information and communication technology (ICT) has to guarantee secure data storage, together with their prompt retrieval and traceability. Moreover, changes in software programmes and operating systems often make obsolete the format of data, limiting and even hindering their access. Data migration due to changes in software tools should be accurately planned in order to preserve integrity of data and documents. It is sometimes advisable to make a PDF copy of main documents, if the translation to a new informatics environment is not likely to be free of errors. Storage and data accessibility must be guaranteed for every researcher or institution interested in them. The UCL Code of Conduct in Research requires that data stored should be checked periodically to ensure that they remain accessible, in case they should be consulted.

The security of data should be considered from three points of view: (1) protection of data and hardware against access by unauthorised personnel; (2) cyber security, i.e. the protection from hackers; and (3) information assurance, namely the act of ensuring that data is not lost when critical issues arise. In order to protect the research data stored in computer systems, the research institution management should consider two specific preventive measures, namely *business continuity* and *disaster recovery*, as suggested by the RQA guidelines [6]. Business continuity is referred to as the discipline regarding the improvement of resilience of an organisation, making it able to maintain operations even when confronting a great disaster. Disaster recovery, on the other hand, implies the ability to quickly and effectively recover to an operational state from a major incident. The business continuity approach enables researchers to maintain access to data in case of computer system failure, while the disaster recovery ensures that, in case of a computer system breakdown, data can be retrieved.

In most cases, adequate skills are not available in the research group and the laboratory relies on external experts or on a dedicated service inside the research institution. Simple rules to be followed in the day-to-day activities are that the computer must be protected by a password, and be backed up on a regular basis. The periodic backup should also be stored at a different site or building, to prevent data loss in case of major accidents.

A final word should be addressed to the problem of the validation of software programmes, spreadsheet and/or computer tools employed to elaborate scientific data, to ensure that all computations are correct and no bias is added to the analysis. Information technology (IT) validation can be guaranteed by the supplier—if any—or performed by the user itself. For example, a simple validation of a spreadsheet can be made by executing the same calculations with a different tool or with a different algorithm.

4.5.5   Validation

The scientific method guides the researcher in identifying the appropriate controls in their experiments, in order to ensure objective, reliable, verifiable and therefore acceptable results. Nevertheless, it must be accompanied by the use of proven methods and does not exempt the researcher from checks on the consistency of data and results.

Validation should be a systematic process, well defined before starting and well documented during the course of the experiment.

Box 4.10: Reorienting the Scientific Method

When crises like this issue of reproducibility come along, it's a good opportunity to advance our scientific tools, says Robert MacCoun, a social scientist at Stanford. That has happened before, when scientists in the mid-twentieth century realised that experimenters and subjects often unconsciously changed their behaviour to match expectations. From that insight, the double-blind standard was born.

People forget that when we talk about the scientific method, we don't mean a finished product, says Saul Perlmutter, an astrophysicist at the University of California, Berkeley. *Science is an ongoing race between our inventing ways to fool ourselves, and our inventing ways to avoid fooling ourselves.*

How scientists fool themselves—and how they can stop
Regina Nuzzo,
Natures News and Comments, 7 October 2015

4.5.5.1 Protocols and Assay Validation

As previously stated, validated protocols are any written experimental procedures, whose results, checked under different conditions and for the agreed necessary number of times, are within an accepted tolerance. Non-validated methods generate non-trustable data, which can seriously harm the research outcomes and, ultimately, the reputation of the scholars. Also, unreliable methods may force researchers to repeat several times the experiments, wasting time and human and material resources.

Often, researchers make use of published protocols and methods, which are considered inherently validated. However, before starting any experiment using a published method, the chosen method should be trialled under the operating conditions of the laboratory where the experiment will be carried out. This should increase confidence in obtaining results with the same level of reliability declared in the literature. On the other hand, when a method is either quite new or derived from another, the validation before use is mandatory to generate credible research data.

The validation of an experimental method in non-regulated research should be tailored to the complexity, scope and the aim of the procedure itself. As suggested by the RQA guidelines [6], more care will be devoted to methods that will be used in key patent or other business-critical submissions, while a lighter approach may be envisaged for early exploratory settings. For the latter, some simple validation methods can be used, for example use of certified reference materials, comparison with other methods and inter-laboratory comparisons. For an extremely rigorous approach, the parameters to be taken into account and verified are specificity, accuracy, precision, linearity, range, limit of detection, limit of determination and strength. This must be accompanied by a sound knowledge and a proper use of statistical methods. Further directions can be found in the ISO17025 and ISO 15189 standards for example, and in several documents of US-FDA and ICH [15].

The RQA Guidelines [6] suggest a validation procedure in seven steps:

1. Initial considerations: Consider the scope of the method and identify regulation, laws and standards for it. Clearly define the purpose of the method and assure good statistical support from the beginning of validation, to ensure a good setting.
2. Method selection and feasibility: Identify/chose staff skill, materials and reference standards.
3. Assay/method development and standardisation: Identify and optimise key parameters, and define suitability controls for both the method and the preparation of samples.
4. Assay/method performance characterisation and validation: Run the method with reference materials and samples, by different scientists, on different days and with sufficient replicates to evaluate accuracy, precision and other goodness parameters. Pay attention to border conditions (e.g. temperature or humidity).
5. Assay/method maintenance and refinement: The method should be monitored to intercept and correct any deviation, from the first symptom of defect.
6. Method verification: A new, even partial validation should be performed whether it undergoes some changes, for example in equipment or materials used, if the method is not used for a period or if it is transferred to another laboratory.
7. Validation report/documentation: Method, conduct and outcomes of the process of validation should be documented in a validation report and the validated method should be described in a procedure distributed to staff, in such a way as to guarantee that everyone is using the same version.

The design of experiment (DoE[18]) can help in defining, planning, executing and interpreting results within the framework of a defined statistical methodology.

4.5.5.2 Result Validation

The validation of experimental and study results can be done by means of two kinds of controls: the first one, regarding the scientific content, requires knowledge in the field of the specific research and scientific experience. It entails controlling the raw data, their processing and the correctness of the conclusions. Colleagues, better if

---

[18] See Chap. 5—Design of Experiment.

not directly involved in the experiment, or external experts may help. This check may take place at different points of study development and certainly on the final outcomes. Ideally, these activities are better done by an external auditor. GLP consider the auditing of data (*study-based inspections*) a mandatory activity in order to guarantee the results.

The second kind of result validation implies the verification of the systems and procedures and how well they are followed to generate data. Research outcomes can also be validated by giving special attention to study design, research conduct and documentation. These checks are in the field of system audits. In GLP they are mandatory and called *process-based inspections*.

Finally, let us share a consideration taken from QPBR: *While it is unusual to implement systematic independent auditing in research institutions, it is only through care at the level of data acquisition and data recording that the study can be validated at all by audit*. In other words, running a controlled process ensures good results that withstand any kind of audit.

4.5.6   Documentation

A good way to introduce the importance of documentation is to underline that it is the *store of collective organisational knowledge* regarding the processes of an organisation. Documentation is the basis for good communication, and allows the sharing of information regarding goals, targets and action plans. Recording procedures and information in an actionable way ensures the alignment of the whole organisation to the general rules, and consequently the results' conformity to requirements, with assured repeatability and traceability. In the view of a management system, documents provide objective evidence of how, when and by whom actions are carried out and to which extent they are compliant. Last but not least, documentation provides a corpus to be used for training of new staff. Most quality standards emphasise that documentation should help the operation with value-adding information, and not be merely the answer to a requirement. For this reason, documentation has to be constantly updated and integrated with any information, modification or new approach to activities, to become a living support to the daily work.

The QPBR makes a clear and useful distinction between prescriptive and descriptive documents, as shown in Fig. 4.10, overtaking and enlarging the former ISO9001 definitions of *procedure* and *record*, respectively.

Prescriptive documents are those providing direction for planning, executing, recording and controlling the activities. They may include technical and organisational procedures (standard operating procedures, SOP), written instructions such as protocols and methods, research proposals and study definitions and planning. On the other hand, descriptive documents keep trace of data, results, observations and decisions. They may include raw data, any derived data and records, study reports and publications.

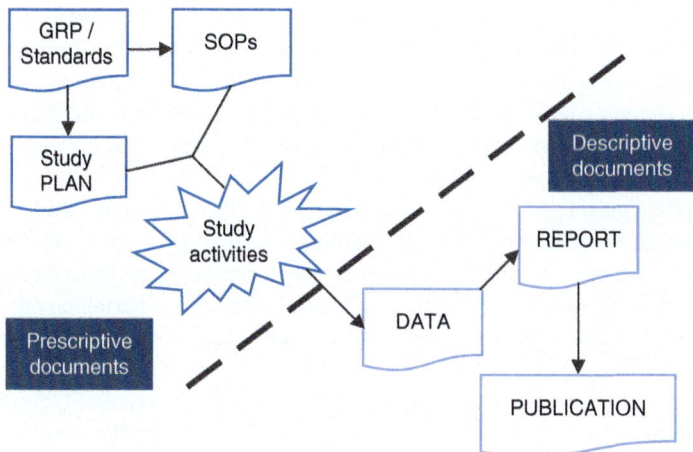

Fig. 4.10 Prescriptive documents, study activities and descriptive documents. *GRP* and other *standards* give directions for drafting *SOPs* and *study plan*, which are *prescriptive documents* and govern the *study activities*. During the execution of the study, *descriptive documents* collect *data* and information in records such as *reports* and *publications* (adapted from [1])

4.5.6.1 Prescriptive Documents

Prescriptive documents can be divided into two main categories, the first concerning all documents regarding general information and directions (e.g. procedures for writing a study plan, for purchasing or for human resource management), and the second concerning all documents giving information and directions specific to a study (e.g. the study plan and assay protocols). For each of them a common template is advisable, as a framework for mandatory contents and revision management, with a format that will be familiar to all who will access, write and share information and procedures. There should be a formal process defining directions for writing, reviewing by relevant staff, approval by management, distribution, update and withdrawal of documents, as well as archiving of obsolete documents that should nevertheless be kept for further reference. This last requirement becomes evident if we imagine reviewing an old study without knowing the rules in place at that time.

All prescriptive documents should be available to all personnel who have to consider them as a reference. To disseminate this information either formal training (mainly for generic prescriptive documents) or a presentation (for the study documents) might be necessary. Staff should follow both generic and specific prescriptive documents and, in case of deviations, these should be documented in the research records with a signed and dated detailed description, so as to preserve the credibility of the system.

Standard Operating Procedure (SOP). Standard operating procedures (SOPs) should be written by the person most familiar with the matter; they describe how an organisational process or a repetitive activity is to be executed. Some examples of organisational processes are how to manage documentation, equipment and staff, or how to write a study plan. Repetitive activities refer mainly to scientific or technical methods: how to perform a protocol or how to operate a specific instrument.

Good laboratory practices (GLP) require writing SOPs for at least the following: test and reference items; apparatus, materials and reagents; computerised systems; record-keeping, reporting, storage and retrieval; test system (where appropriate); and quality assurance procedures.

One of the most important aims of a system of SOPs is to give researchers a clear and unique pathway for major research activities, so that the research process can be easily reconstructed. In this light, research methods should be described in sufficient detail, with special regard to exclusions and exceptions. Particular care should be devoted to statistical treatment: elaboration and analysis of data should be clearly defined.

The QPBR provides a list of benefits in maintaining an organic SOP system, as well as some hints to succeed in enforcing the application of SOPs. Among the first benefits mentioned are the opportunity to optimise processes; easier documenting of complicated techniques in study plans; a means of study reconstruction; and a means of communication in case of audits, client and/or sponsor visits and technology transfer. Towards the end of the list but equally important are assuring the support and commitment of upper management, and maintaining constant SOP training of staff.

Research Proposal and Study Plan. The research proposal is the document by which a research group describes the scientific context, overall objectives and scope of a research programme, to apply for a grant. Funding agencies usually provide their own form to collect and organise the information required by the grant scheme, so it is not worth to delve deeper into this subject. This document is the basis for the most important planning document of any research project, i.e. the study plan, and, for this reason, it has to be well written and then properly stored. Conversely, the *study plan is the key document for communicating the intentions of the study to all contributing staff and sponsors* (QPBR).

It would be useful for the research institution to have guidelines (or SOPs) for the drafting of both research proposal and study plan, detailing how they are to be written, reviewed, approved and amended.

In addition to the proposal contents—namely, scope, objectives, time frame, milestones and deliverables, resources and institution/laboratories involved—the study plan should contain all details necessary to drive unambiguously the research project. Taking inspiration from GLPs, the contents suggested are:

- Identification of the study
- Identification of the research institution(s) and related facilities, study director and researchers
- Dates of approvals, reviews and start and end dates of the project
- Scientific methods
- Detailed information on the scientific project
- Documents to be issued

Materials and methods should be carefully described first to give proper direction to staff and second to allow reconstruction of the work in adequate detail. The QPBR should be of help [1], providing a useful list of the information to be included in this section, comprising test material, conditions for handling and storage, type of products, equipment and information about data handling, reporting of results and, if applicable, reference to other works. A paper by T. M. Annesley [16] suggests that, in order to structure in the best and complete way the section Materials and Methods of a scientific paper, the simple questions of *Who, What, When, Where, How and Why* should be answered. This technique is known as *5W and 1H* or *The Kipling's Servants* and will be described in Chap. 4.

The study plan is approved by the study director, as appointed by the research institution's upper management. They then have the responsibility to conduct the study accordingly, as well as to record and properly justify any minor deviation and obtain authorisation for major ones.

4.5.6.2 Descriptive Documents

At the core of the research project are the research records, data, annotations, elaborations and conclusions. Any publication, patent application or further research project must be based firmly on them, to avoid waste of resources, time and funding. As highlighted also by the QPBR, scientific interpretation, as well as any publication that may follow, stands on the basis of the records made of scientific activity results by responsible scientists, and become the basis for any subsequent publication. The data, collected and stored in a proper way, represent the value of the research in terms of time and resources employed in the studies, as well as their economic potential.

Research and Work Records. Research and work records are probably the most important aspects that, if mismanaged, lead to scarce reproducibility of research results (see Introduction). As clearly stated in the RQA Guidelines *during the course of research many (…) records will be generated and they are often the critical pieces of the puzzle—what was done differently that day, what was the new approach tried, who was the scientist who made the discovery? Efforts should be made to retain these support records so valuable information is not lost, otherwise work may need to be repeated, or key conclusions may be missed* [6].

Clear rules for data identification, collection and storage should be established before the research project starts, to ensure that data are properly preserved and their retrieval and traceability are secure. In addition, cross-referencing among different records and documents should be carefully described and governed. The record and document management should be described in a written procedure document or follow a well-established formula.

Collecting Data. Raw data, namely the original data collected through measurements, instruments or by other means, are part of the information about the research that needs to be recorded. The FDA in its guides to GLP defines *raw data* as *any laboratory worksheets, records, memorandum, notes that are the result of original observations and activities and are necessary for the reconstruction and evaluation of the report of that study* [17, 18]. Some examples are:

- Records/forms
- Instrumentation printed outputs
- Labels (identifications of animals or materials)
- Spreadsheet/worksheet/laboratory notebooks
- Calibration certificates
- Evidence of education and training

In the following, requirements of GLP and patent applications are briefly illustrated. These requirements may seem an excessive burden for the scope of certain experiments in non-regulated research; in this case the researcher may draw inspiration from GPL while keeping the discretion not to apply all of them. For example, the strict rules for raw data may not be completely followed, but the researcher should find the right way to avoid loss of data, errors or ambiguity. A good way to decide which raw data are to be recorded and when is to perform a risk analysis, which helps in identifying the most probable problems that may arise during the study, assessing the related risk and defining actions to limit or reduce them.

GLP give accurate direction on how to record and maintain traceability of raw data [19]. Raw data records must be written in ink (no pencil notes allowed), and the data must be:

- Directly, promptly and completely recorded by the person performing the work, on a permanent support (not on loose sheets); in case of patent application, the signature of a supervisor may be required
- Clearly readable, no doubts or queries possible
- Accurate
- Signed: name in capital letter, signature, date in a non-ambiguous format
- Identified according to a dedicated procedure

Moreover, corrections made on data or other details must not obscure original data, and again must be dated, commented and signed.

Another type of data to be recorded is represented by the experimental conditions and the background: protocols and controls; quantity, characteristics and batch number of materials and reagents; environmental conditions, if pertinent; in short, every piece of information useful to replicate the experiment or to deepen the investigation.

The researchers must make use of notebooks to record information during the conduct of experiments. Notebooks can be either conventional, paper logbooks or electronic ones, depending on the inclination to innovation of the research team. These latter devices are known as electronic laboratory notebook (ELN) and are available in a wide variety of type, size and performance, both free and commercially available. They enable the researcher to store, organise and publish the research data, at the same time maintaining their integrity, traceability and security. Using an ELN streamlines some operations on data, e.g. searching, sharing and linking to instrument record. The ELN may also allow managing changes in data according to traceability requirements. A laboratory notebook can be used to record information about a single study, the work of an individual researcher or all the activities in a laboratory, with the choice resting on the laboratory or institution management. It is strongly advisable to dedicate one laboratory notebook to each study, avoiding both the mixing of different studies within the same book or scattering data of the same study in the books of different researchers.

These are some hints for the correct management of laboratory notebooks for non-regulated research, inspired by the indications of QBPR and GLP:

- All notebooks of the laboratory should be numbered consecutively, so that they are easily identified.
- All pages of each notebook should be numbered before they go into use, and never torn out.
- Some blank pages should be left at the beginning of the book allowing the creation of an index.
- Information regarding any activity and related data—for example, samples for assay, and the assay results—should be recorded and/or referenced, including any data or specimens held in another location as well as computer files.
- All information recorded should be signed, possibly at the end of each working day.
- Corrections should be made in a timely manner and in such way that the original entry remains visible, recording the reason for it and putting signature and date of modification. Computerised data and information should only be modified by means of a digital signature system that allows to record identity, date and motivation of change.
- Notebooks should be kept in a safe place when in use and promptly stored when completed.

Elaborating Data. There are two remarkable hazards to the correctness of elaborated data: inadequate staff skills and misuse of software programmes. In short, to ensure the reliability of their results, the researchers have to not only employ tools and software programmes that are proven to be error free, but also be skilled as to their use and their methods of application.

A typical example is the statistical elaboration of data. Statistics is a complex discipline that has to be fully mastered in order not to draw false conclusions, thus compromising the research results and subsequent decisions. It is incredibly easy to misunderstand the statistical significance of a set of data, on account of an experiment not being well dimensioned or data not properly collected. A statistical method useful to plan and execute experiments, as well as to draw out more information and to achieve a good statistical robustness, will be illustrated in Chap. 5. Data elaboration (statistics) competence is one aspect of the staff information that has to be, in the first instance, assessed and registered, then monitored and periodically updated.[19] Such competence may also be sought outside the research team. In that case, statisticians are the *critical suppliers*: they should be carefully chosen and

---

[19] See Sect. 4.5.3.2.

evaluated. Moreover, the analysis should be thoroughly discussed between the research team and the experts, to ensure the correct interpretation of data and results. A similar approach applies to the skills for the use of sophisticated equipment employed in data collection and preliminary elaboration, and should be programmed by the user.

Sometimes a spreadsheet or a short software programme built by the researcher themselves may be used to elaborate data. Those kind of home-made tools are quick and easy to use, but can easily hide computational errors. Thus, they have to be validated before being employed to generate results. The simplest way to validate *self-made* software tool before the first use of the spreadsheet or software programme is, in the first instance, to compute the same data with a different algorithm to ensure the correctness of the calculation. This precaution does not apply to commercial software programmes, although a critical glance at the first batch of results is always an advisable practice, not least to ensure that the software programme has been correctly used. To guarantee that data and their elaboration can be tracked back, all versions of the software tool used should be recorded in the laboratory notebook, along with references to data and any manual calculations performed.

Publishing Data. It is worth emphasising two aspects of the publication of data. The first aspect is the dissemination of all the data from a project to nurture the growth of human knowledge. In particular, data obtained with public funding should be shared. As stated by the Research Councils UK (RCUK), *Publicly funded research data are a public good, produced in the public interest, which should be made openly available with as few restrictions as possible in a timely and responsible manner*. The RCUK issued in 2015 and updated in 2017 a policy for the sharing of research data [10], which addresses the main implications of the issue, such as the ethical, legal or business constraints, and the need for the researcher to maintain secrecy at least until the their work has been published, as well as how to optimise the use of public research funds. A guideline is attached to the policy to help in the drafting of an actual data management plan. Although the policy is specific to the councils affiliated with the RCUK, it can be a source of inspiration for those who wish to tackle this issue in a comprehensive way. A summary of the principles of the RCUK Data Policy is illustrated below.

The second aspect concerns the selection of data to be published, which should not be based on goodness, but on truthfulness. In the class of truthful information negative results are also included, even though they are often excluded from the publications because they are unprofitable in terms of image and prestige. Negative results however are valuable information that enriches the knowledge, allowing others not to take paths already travelled without success, thus saving resources for new lines of research. As so eloquently expressed by Albert Einstein, *By academic freedom I understand the right to search for truth and to publish and teach what one holds to be true. This right implies also a duty; one must not conceal any part of what one has recognised to be true.*[20]

---

[20] Albert Einstein, Letter on his 75th birthday, 1954, reported on his Memorial Statue at the National Academy of Science, Washington DC, USA.

4.5.6.3 Study Report

On completion of the study, all raw data, elaborations, results as well as product identification, protocols and study plan should be collected in a *study file*. Besides publications and/or patent application, this will allow a complete and correct reconstruction of all research work and results.

GLPs require a minimum content for the (mandatory) final report. In addition to the information that identifies the study, the team and its managers, and any parts of the study entrusted to third parties, the report should also include the following:

- Test and reference item and their characterisation
- Sponsor and test facilities
- Start and completion dates
- Materials and methods
- Results including calculations and determinations of statistical significance
- Discussion of the results and conclusions
- Storage of study plan, samples of test and reference items, specimens, raw data and final report

In addition to this, it is suggested to record equipment and software programmes used in the study, because they could significantly affect the results and are essential for reconstructing the study. Incidents and deviations should be reported as well. A list of references should be included, to both laboratory notebooks and literature.

It is essential to report any kind of results, both positive and negative or conflicting. As stated by the QPBR, *Leaving out results that do not fit is called selective reporting and is not an acceptable practice. Results that do not fit may be precisely those that lead to an improved hypothesis for further testing.*

In any case, the principal investigator is in charge of checking and takes responsibility for the final report contents: information, results and conclusions. Again, in the non-regulated research the kind and detail of the information to be recorded in the final report can be defined according to the scope and aim of the study, as well as the requirements for tracing and reconstructing data and results.

4.5.7   Storage and Archiving

Documents and samples produced by a research project are a scientific heritage surely worth preservation, but they will be useful for the future only if storage and archiving have been done properly, with appropriate identification and clear directions for conservation. Otherwise, in a research environment subject to high staff turnover and heavy intellectual commitment, it will be difficult to find not only what is needed, but also to reconstruct what was achieved, in other words to keep extracting value from old research studies. Moreover, samples and documents occupy space, creating the need for adequate and expensive storage devices such as freezers or fireproof cabinets: this habit becomes a burden on future budgets of research institutions and projects.

4.5.7.1 Data Archiving

It is not the aim of this book to deal with the problem of *big data*. Some precautions however are suggested to avoid common problems with data storage. Any data record, clearly identified when collected, should be stored in such a way to streamline consultation and retrieval. During the course of the study, dedicated areas (both physical and electronic) may be identified to keep all records, communications and documents, before the formal report is completed.

Some regulations, as GLP, require that the modifications to raw data are tracked, noted and verified by a responsible person. This precaution should be adopted also in non-regulated research, to improve traceability of data, including computer-stored data, with appropriate software programmes. An electronic laboratory notebook (ELN) can easily deal with this requirement. Moreover, it would be advisable to charge only one person with the supervision of the study data chain, to avoid differences in management that may cause misalignments. For the same reason, a standard operating procedure may be very useful.

Again, taking the GLPs as a guiding light, archiving should be done at least on the study plan, raw data and final report. GLPs also require maintaining records regarding staff training, qualifications and skills, validation reports, maintenance and calibration records of equipment and computer systems, and environmental monitoring records. All these documents are important also in non-regulated research to maintain control of the research environment. Furthermore, research institutions should explicitly require corresponding author/s of a paper to be submitted for publication to have under strict control all the information that led to the generation of the data presented in the manuscript, with specific attention to those used for producing figures and tables. The observance of this guideline, on the one hand, should make explicit the accountability of the PI for the integrity and reproducibility of data and, on the other hand, should greatly facilitate modifications of the paper in response to reviewer comments.

Both QPBR and GLPs indicate that access to the archive should be limited to authorised personnel, and that only one person should be charged with the management of the archive. The archive should be protected against major damage. Useful ways to ensure protection are fire cabinets for paper documents, and periodic back-ups stored in a different location for electronic documents. The records should be preserved for at least 5 years after the publication, unless otherwise specified by regional or institutional regulation.

4.5.7.2 Sample Storage

Storage of samples should be carried out in order to guarantee their integrity, traceability and conservation at all times. All significant materials used and produced during a study should be clearly identified and stored for a period depending on the duration of the study, regional laws and/or research institution regulations. The identification includes details about methods/protocols used. If the storage conditions are critical—temperature, humidity, etc.—these have to be properly monitored and the corresponding records properly kept.

GLPs identify and require an archivist role who is in charge of storage and archiving of samples and documents. The position of archivist, requiring specific but not massive training, is a good solution in an institution of a certain size/magnitude, where these tasks drain time and effort away from primary research activities.

4.5.8   Health and Safety

The use of procedures and materials that are potentially hazardous must be adequately set to guarantee the safety of staff and facilities, as well as of the environment. The principal scientist, or a person appointed by the research institution, must inform all staff how to properly use hazardous materials, equipment and procedures and to comply with the health and safety regulations, safety at work and environmental protection measures in place. In every research institution, international, national and local, regulation must be made available to researchers and staff. The availability of these documents is also part of the mandatory requirements of an ISO 9001 QMS.

4.5.9   Publishing Practices

Publishing the results of a study is the natural conclusion of a research project, and it has better impact if done in a timely manner because it allows the researcher to have feedback from reviewers and colleagues, and for the results to be immediately disseminated to contribute to the growth of knowledge.

Research organisations should have a policy for publishing results and this policy should include main directives for the time of publishing, the procedures for the review and release of the work, the authoring conventions, and the attribution of intellectual property and ownership of patent rights (between researcher, research institute or grant-giving body), as also suggested by the QPBR.

The decision to publish original results of a study in a single paper or to make separate publications for different details is incumbent on the principal scientist and the co-authors. Also review papers, summarising different studies within the same scope, are very useful for other scientists, even if this route should not be overexploited to enhance the list of publications.

Scientists are still reluctant to publish negative results, but if they are obtained from a properly managed study, then they are very useful to drive further research, enrich knowledge on the specific subject and avoid wasting of time and resources lest other scientist should want to follow the same path.

Defining the list of authors is a delicate subject [20]. An unwritten international convention indicates that the position of names in the list of authors refers to the order of the relative responsibilities for the content of the paper. The list of authors should include only the researchers actually giving a significant contribution to the work. The QPBR condemns the habit of inserting the head of the institution and/or researchers with a very marginal role, because the responsibility for data, results and conclusion becomes too scattered among them.

The choice of forum in which to publish is crucial for giving the due importance to the work. Best choice would be a peer-reviewed journal with an adequate impact factor (IF) that attests to a good reputation and importance, but publications at various stages of development in the research project may be submitted to conferences as posters or oral presentations. Sometimes researchers are asked to contribute to the popular press for example articles in non-scientific journals, conferences and exhibitions. In this case, the scientific matter and language are simplified to meet popular knowledge and the only attention suggested is to ensure that contents can be released into the public domain.

The institution should have also a policy for dealing with conflict between publications and patent application; this matter is, however, outside the scope of this book.

4.5.10 Ethics

Box 4.11 shows the opening statements of the European Code of Conduct for Research Integrity, issued in 2011 by the European Science Foundation (ESF) and All European Academies (ALLEA), which stresses the importance of defining norms and rules to assure the ethical conduct of the scientific research. Ethics in scientific research may refer to different domains, regarding the safety of laboratory, the impact on the environment and the subjects involved in the study, adequate training and skill management of the researchers involved in the project, correct management of study data and the conduct of the research itself, ending with a proper publication of results. Ethical publishing has already been dealt with in the previous section. As stated by the QPBR, the principal investigator is deemed responsible for any negative effects on people (both research staff and community), animals and environment.

The involvement of laboratory animals in research is taken into consideration by the good laboratory practice (GLP), complemented and supported by national and supranational regulations aimed at preserving the ethical use of animals for scientific purposes. The QPBR suggests to carefully assess the number of animals required for the study: neither too many, to avoid useless sacrifice, nor too few, a condition that would undermine the statistical significance of the results. Moreover, many research organisations voluntarily adhere to the directive of the *Association for the Assessment and Accreditation of Laboratory Animal Care*—International (AAALAC), a non-profit organisation gathering, through the institution of accreditation, more than 650 research entities throughout the world.

The QPBR, among other recommendations, points to respecting biosafety, namely the impact that creating and working on new species of organisms could have on people and the environment, and the need for scientists to comply with national and international regulations or, if these are not established, to be guided at least with the protocols in the WHO Laboratory Biosafety Manual [21].

The QPBR also invites scientist with a responsible role to perform a risk analysis before starting a study, preferably integrating it in the application for funding, in order to identify the hazards and limit the possibility of mishaps. Addressing this aspect correctly should include training staff about safety, safe behaviours and good animal welfare.

It is strongly advisable that every research institution issue a code of conduct, to maintain an adequate and uniform standard of ethics. The University College of London (UCL) has issued its *Code of Conduct for Research* in 2013 [22], focusing on professional integrity of researchers, design and methodologies for research, publishing and ensuing responsibilities of the leadership and the institution. Another meaningful document is the one issued by Research Councils UK (RCUK) [10], where sections are dedicated to a classification of unacceptable conduct and to the procedure for reporting and investigating misconduct. In detail, unacceptable conduct is listed as fabrication, falsification, plagiarism, misrepresentation, breach of duty of care (whether deliberately, recklessly or by gross negligence) and improper dealing with allegations of misconduct. The University of Cambridge, among several institutional initiatives intended to promote research integrity, released a leaflet that, in two pages, summarises directions on ethical research conduct. It is very good practice to spread the information and the directions widely throughout the institution [23].

Box 4.11: The Importance of Research Integrity
Science is expected to enlarge mankind's knowledge base, provide answers to global challenges and guide decisions that shape our societies. Yet when science is compromised by fraudulent activities, not only the research enterprise stumbles, but also society's trust in it. Thus, researchers and leaders throughout the world should ensure that science is trustworthy to our best knowledge. This can be achieved by education, promoting a culture of integrity, and by development of and compliance with joint rules and norms.

From the foreword of the former edition of *The European Code of Conduct for Research Integrity*,
ESF/ALLEA

4.6   Examples of Application

4.6.1   Managing Research Resources in a Large Institute: The San Raffaele Experience (Daniele Zacchetti[21])

The San Raffaele Scientific Institute in Milano is the largest private research institute in Italy, with about 100 independent research groups, 25 secretariat offices and

---

[21] San Raffaele Scientific Institute, Italy.

almost 1000 people working in basic and preclinical research areas (including tenured scientists, postdocs, PhD students/doctorate candidates, undergraduate students, technicians) for an area of over 130,000 m². The institute produced 1381 scientific publications in 2016 alone.

Those numbers are an obvious challenge for the proper management of resources: consumables, reagents and instrumentations.

Researchers raise research funds by applying to public and private institutions, mainly charities. The administrative resources of the institutions that constitute the San Raffaele Scientific Institute (the San Raffaele Hospital, the San Raffaele Foundation and the San Raffaele University) manage these funds, but researchers have a high degree of autonomy in the way the funds are expended. This is a possible source of inefficiency if orders of reagents and consumables, storage of goods or use of instruments are not properly governed. In fact, a laboratory may place orders without taking into consideration similar orders by other laboratories; there might be multiple stocks of the same reagents and instruments might be underused or incorrectly distributed throughout the institute. Since only a minority of researchers hold a tenured position, there is a high turnover of people (especially graduate and undergraduate students, and postdocs), creating a condition that favours long-term storage of reagent, even beyond their expiry dates. Moreover, instruments may suffer damage due to lack of maintenance and, more frequently, to improper use or inexperience.

All these issues should be tackled by a comprehensive approach based on centralised procedures, kept under the control of the Scientific Directorate, with the support of informatics tools. The software support for these activities can be often found in the community of freelance/open-source developers (traditionally very active in the scientific field), thereby reducing the costs.

In San Raffaele Scientific Institute, an office, led by a scientist (taking advantage of his direct research experience), has been recently established with the aim to optimise the use of resources, thereby sparing research funds by coordinating orders, sharing reagents, rationalising storage spaces and optimising the use of instruments. As part of the internal San Raffaele policy, high-tech and expensive instruments are centralised in dedicated facilities; this is the case in the areas of advanced microscopy, genomics and proteomics, while other, more commonly employed, instruments are relocated according to need but are still monitored by a Web-based booking software that allows reporting. When an instrument is not managed by a facility at least one person is in charge of it (usually a researcher who frequently uses the instrument and has a vested interest in maintaining the resource in good shape), and who has also the responsibility of training new users. Regarding reagents, a centralised and uniform system transmitting orders from the secretariat offices to the different administrations is under development in order to help researchers to trace similar orders with the date and the name of the person who placed them, thereby allowing direct contact to obtain information about the product and possibly test the reagent or share it. A new centralised system monitoring the storage and the shelf life of laboratory reagents is also under evaluation.

Overall, proper management of the research resources helps not only in keeping costs down through optimisation of the existing resources and reduction of waste, but also in containing the dreadful costs of flawed biomedical research caused by poor reproducibility of scientific results [24]. An estimation of 20% savings through correct management of the available resources is not far from reality and would allow scientists to increase efficiency and get the best value for the precious grant moneys.

4.6.2   A QMS for a Life Sciences Research Laboratory and Its Related Management Software (the qPMO WP3 Team[22])

In the context of the qPMO Network,[23] the Working Package 3 (WP3) created a QMS model to translate the methods and skills related to quality in industry and apply them to research [25]. The development, optimisation and validation of the system, inspired by quality and project management (PM) principles, are considered as an innovative and simplified way of planning and organising research activities, overcoming the widespread prejudice in the scientific research environment that they are an impediment to creativity.

The Life Sciences QMS Model—To create the QMS model it was necessary to translate the methods and skills relative to quality in industry and apply them in research. From a range of available models (i.e. GLP, ISO17025), the international standard ISO 9001 was selected and applied since it is the most popular among the standards that allow a system certification. First, a pilot laboratory was selected: a research laboratory of the C.N.R., working with marine animal models (mainly the sea urchin *Paracentrotus lividus*) in the area of drug discovery and embryonic development. The operational and support processes to be managed were selected, and then the primary processes were placed under quality control: research activity, student training and science communication. The main secondary processes impacting them are processes for managing resources (e.g. provision, personnel and instrument management, economic resources) and quality management (e.g. verification and improvement, management of records, internal and external communication), as illustrated in Fig. 4.11. Stakeholders, recipients and suppliers for the research laboratory were identified and, finally, procedures, operating instructions, guidelines, laboratory notebook formats and layouts to cover all laboratory-related processes were generated.

The QMS model was created, applied and validated for the pilot laboratory with the final aim to generate a total quality management (TQM) model easily transferrable to other research laboratories and finally to scientific public institution. The laboratory QMS is certified according to ISO 9001:2008 and is currently under revision to be updated to the 2015 edition of the standard. A proof of concept of QMS efficacy in the pioneering laboratory has been derived from the evaluation of

---

[22] Antonella Bongiovanni, Marta Di Carlo and Luca Caruana.

[23] See Introduction–Acknowledgements..

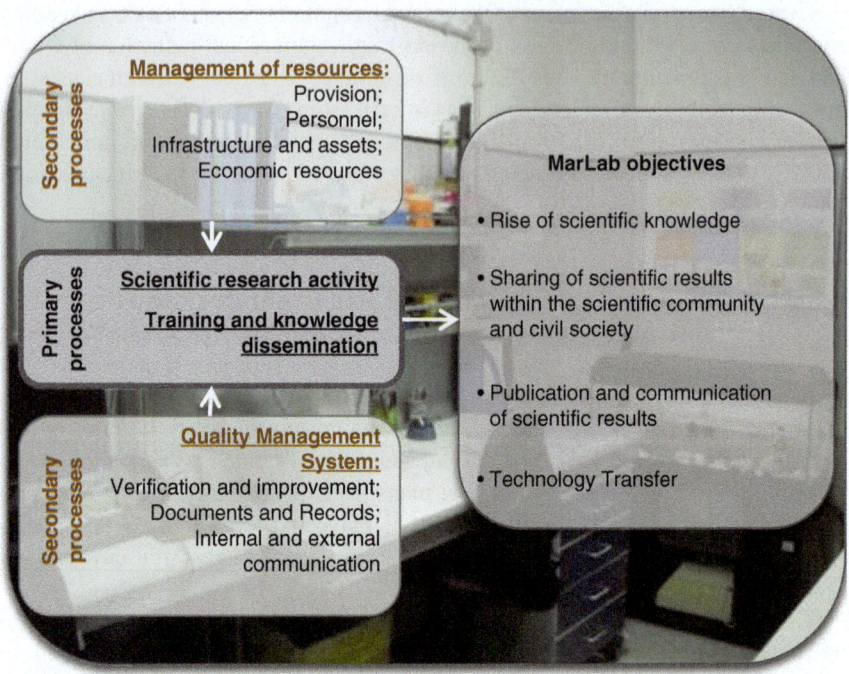

Fig. 4.11 Processes related to the activity of a research laboratory. In the MarLab QMS, primary processes are *scientific research activities* and *training and knowledge dissemination*. Secondary processes are grouped into two classes: processes for managing resources and processes for managing the QMS, both supporting the primary processes, in order to achieve the MarLab objectives

selected quantitative indicators (i.e. staff motivation rating, number of nonconformity, efficiency of aquarium management and number of published scientific papers) 3 years after certification, during the internal audit. As far as the management of the sea urchin housing in the aquarium was concerned, the related indicator was found to be *excellent*, ever since the first year of QMS application. Moreover, since the application of QMS, none of the programmed experiments had to be postponed or cancelled due to lack of suitable biological material, namely the sea urchin embryos, thus confirming the good management of the aquarium under this management system.

The Life Sciences QMS Management Software, Help4Lab—Based on the QMS model, the Help4Lab modular software programme was developed to manage quality, safety, equipment, environment and documents in a research laboratory. Help4Lab has restricted access: any user with the username and password can navigate within it, limited to their role, while technical and quality managers and scientific director have full access. Help4Lab includes a section *Processes* that supports the management of all documentation (management procedures, guidelines, operating instructions and forms) inherent to processes identified in the research laboratory. A second section *Provision* manages the entire path of the product, starting from a list of suppliers which includes the companies involved, contacts, type of product and, if available, certifications. Each supplier is subject to evaluation, dictated by specific variables that determine the reliability of the sales service and the quality of the product. Once purchased, products, as well as their consumption, are recorded and catalogued; thus the stocked quantities are continuously updated. To avoid being suddenly devoid of a product due to a higher, unexpected consumption, a *threshold* quantity (minimum stock) was determined, below which the system sent an alert to reorder the product. In the *Instruments* section, each instrument is recorded and catalogued together with information regarding its maintenance. In the summary data windows, the good operating ranges are highlighted, so as to provide the operator with an immediate graphic feedback of the trend. All the maintenance operations are scheduled in a specific and timely manner, with alerts connected to email addresses and phone numbers of the technical manager. In this section all maintenance alerts are identified and processed by the maintenance event diary, alerting the technical manager. Furthermore, the section *Management tools* assists in the planning of maintenance and calibration of instruments. The software programme has been temporarily recorded in the CNR's Intellectual Property Rights Portfolio.

The QMS model and the related management software programme Help4Lab are new tools for improving and simplifying the organisation of research laboratories. This QMS model has motivated the staff towards continuous improvement of shared operations, enhancing communication between all management levels and personnel. Such a system also ensures greater reliability of the results of research laboratories. Furthermore, the QMS model has contributed to increase the prestige of the laboratory and the public research institution.

4.6.3   GLP as a Reference for Non-regulated Research (the MoBiLab Team[24])

A research project, aimed at setting up a laboratory information management system (LIMS) based on good laboratory practice (GLP), for the Molecular Biodiversity Laboratory (MoBiLab) of the research infrastructure LifeWatch (www.lifewatch. eu), will be presented in this section. In fact, taking into account the progressive evolution of GRP, and also considering that the *customers* of research services are laboratories working under the principles of good laboratory practice (GLP), MoBiLab decided to voluntarily adopt an internal quality system, establishing a connection between GRP and GLP (i.e. from non-regulated to regulated research) and paving the way for an ISO 9001 certification.

Good laboratory practices (GLPs) are mandatory in OECD countries for preclinical tests, but may also be considered, outside their main scope, as a reference for laboratory management systems that can be referred to as *GLP-like* quality

---

[24] Francesca De Leo, Marinella Marzano and Caterina Manzari.

systems. Applying a GLP-like quality system allows a research laboratory to maintain under control staff structure (role, accountability and responsibility, skill and competence), research study development, management of substances under testing, test vehicles, materials and equipment, facilities, automated systems and documentation, according to clear and strict rules. Such a management system ensures the quality of the research outcomes: uniformity, consistency, reliability and reproducibility.

In MoBiLab, skills and advanced facilities for molecular and bioinformatics analysis are integrated to provide the scientific community with services and advice for molecular biodiversity studies. The MoBiLab facilities allow to obtain detailed taxonomic and genetic/functional information from environmental, food or clinical samples. The main MoBiLab functions are (a) the design and application of massive meta-barcoding—or shotgun DNA sequencing protocols—for the analysis of prokaryotic and eukaryotic genomes; (b) the profiling of transcriptomes of individual organisms; and (c) the investigation of eukaryotic, prokaryotic and viral microbiomes living in different environments including human hosts.

The laboratory is fully equipped with operative platforms based on the most innovative next-generation sequencing (NGS) technologies and powerful resources for data storage and computational analysis. Next-generation sequencing (NGS) platforms share a common technological feature consisting in the massive parallel sequencing of clonally amplified or single-DNA molecules that are spatially separated in a flow cell. This design is far from the classic Sanger sequencing based on the electrophoretic separation of chain termination products generated in individual sequencing reactions. NGS platforms have radically changed the field of genomics allowing both resequencing and de novo sequencing of whole genomes and are routinely applied to a variety of functional genomics problems, including, but not restricted to, global identification of genomic rearrangements, investigation of epigenetic modifications, single-nucleotide polymorphism (SNP) discovery, transcriptome profiling and metagenomics. Although NGS has markedly accelerated multiple areas of genomics research, it is a massive parallel process and, thus, generates an unprecedented volume of data, presenting challenges and opportunities for data management, storage and, most importantly, analysis and interpretation.

The final goal of the quality management system has been to develop an integrated approach based on good laboratory practices (GLPs) accompanied by a LIMS management system in order to ensure the highest levels of reliability, reproducibility and traceability of the results, a process that is also expected to entourage their potential exploitation. The LIMS is a computer system capable of handling the acquisition stages, the processing and the storage of all data generated by a laboratory and/or processes. With LIMS a laboratory can have a single platform to manage all tasks and thus focus more on the laboratory business without wasting time on paper management, also minimising the risk of errors. Summarily, the security of information and the reduction of errors in data entry and costs represent the main advantages of LIMS.

MoBiLab first outlined a description of the main process by means of a supplier-input-process-output-customer (SIPOC)-like flow chart, which includes the person

in charge for each activity and the relative documented information (standard operation procedures, SOP, and records). After having identified the SOPs necessary for
the operational processes, researchers were provided with a template and with the
instructions to draft them. At the same time, management procedures were defined
by the whole team, drafted and supported by flow charts and other quality tools
whenever necessary. SOPs were ranked with respect to priority; however, a few
SOPs were not considered since they were not required for these specific research
activities (e.g.: management of test systems). Bioinformatic SOPs were written
afterwards in collaboration with expert. The Italian company Eusoft was tasked
with developing and optimising the LIMS platform for managing all the laboratory
activities through a suite of integrated modules, to ensure adequate support to data
and information management.

The platform was structured starting from SIPOC-like flow chart and the modules were developed and customised in agreement to the SOPs. The main function-
alities implemented in the system include the grouping of all samples from the same
project and the analytical steps categorised in the system. The control of the execution of all stages has different tasks: it warns the operator in case of failure of one
phase; the platform automatically acquires the results produced by the instruments
and records raw data; manages the files related to the tests; provides automatic integrated reporting and traceability of instruments and materials used in the analytical
stages; and supports the management of the maintenance schedules of instruments
as well as that of staff roles and qualifications based on the organisation chart. A
corrective action/preventive action (CAPA) system will also be implemented within
the LIMS.

The designed system offers several advantages in the management of laboratories working on genomic and data analysis that represent a great innovation in the
current management of analytical activities:

The system, thanks to its organisational charts, defines the responsibilities of
each team member; the detailed process flow chart ensures sample traceability.
Also, data is automatically transferred and the efficiency and reliability of equipment can be closely monitored. Furthermore, the system allows control of all materials and reagents, of the progress of research activities and of the security aspect of
laboratory management (authorised access). Simplification and reduction in the
number of phases as well as rapid production of reproducible results are other significant benefits of the system (Fig. 4.12).

The implementation of some lean tools in the production process of MoBiLab
will be the next task of the project. Lean management[25] (a.k.a. Toyota Production
System) is a comprehensive methodology developed in manufacturing to reduce
waste and improve product quality. A value stream mapping (VSM) will be outlined
to identify value-added and non-value-added activities, in order to improve the process flow. Workplaces will be optimised by the 5S method, and *poka-yoke* systems
will be introduced for identification, traceability and handling of samples and data.
The lean approach will be strictly connected with the LIMS system, leveraging its

---

[25] For Lean management and related tools, see Chap. 5—Lean Management.

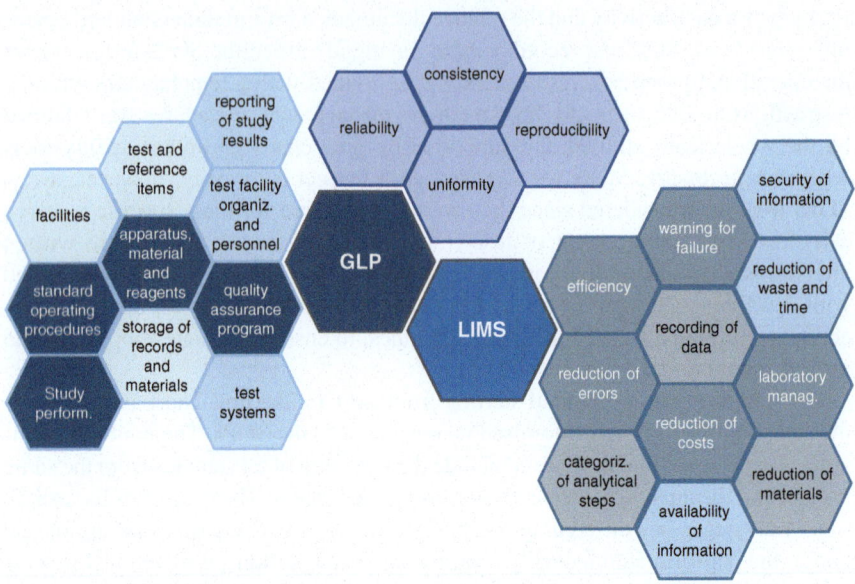

Fig. 4.12 GLP-like quality management system and LIMS. The beehive on the left summarises the features of a GLP-like QMS, while the beehive on the right summarises those of a LIMS. Together, GLP-like QMS supported by a LIMS allow achieving *reliability*, *consistency*, *reproducibility* and *uniformity* of scientific data and results

features to ensure optimal control. The GLP-like system and the lean approach are expected to improve the efficiency of the MoBiLab, limiting the waste of time and materials and reducing opportunities for error while maximising effectiveness, thus guaranteeing traceability and reproducibility of results.

References

1. WHO: WHO-TDR handbook: quality practices in basic biomedical research. 2010. http://www.who.int/tdr/publications/documents/quality_practices.pdf?ua=1. Accessed 13 Sep 2017.
2. A guide to the Project Management Body of Knowledge (PMBOK® guide). Project Management Institute (PMI). 2017.
3. Robins MM, Scarll SJ, Key PE. Quality assurance in research laboratories. Accred Qual Assur. 2006;11:214–23. https://doi.org/10.1007/s00769-006-0129-5.
4. Davies R. Good Research Practice: it is time to do what others think we do. Quasar-RQA. 2013;124:21–3.
5. Pentland A. The new science of building great teams: Harv Bus Rev; 2012.
6. RQA working party on quality in non-regulated research. Guidelines for quality in non-regulated scientific research booklet. RQA. 2008.
7. Eaker S, et al. Concise review: guidance in developing commercializable autologous/patient-specific cell therapy manufacturing. Stem Cells Transl Med. 2013;2(11):871–83.
8. UK Government: joint code of practice for research (JCoPR). 2015. https://www.gov.uk/government/publications/joint-code-of-practice-for-research-jcopr. Accessed 13 Sep 2017.

9. RQA: quality systems workbook. 2013. https://www.therqa.com/assets/js/tiny_mce/plugins/filemanager/files/Publications/RQA_Quality_Systems_Workbook.pdf. Accessed 18 Sep 2017.
10. Research Councils UK: RCUK policy and guidelines on governance of good research conduct. 2017. http://www.rcuk.ac.uk/documents/reviews/grc/rcukpolicyguidelinesgovernancegoodresearchconduct-pdf/. Accessed 13 Sep 2017.
11. ESF: the European code of conduct for research integrity. http://www.esf.org/fileadmin/Public_documents/Publications/Code_Conduct_ResearchIntegrity. Accessed 13 Sep 2017.
12. Universitat de Barcelona: code of good research practices. 2010. http://diposit.ub.edu/dspace/handle/2445/28544. Accessed 13 Sep 2017.
13. Universitat Autònoma de Barcelona: quality policy. 2015. http://www.uab.cat/web/research/itineraries/uab-research/quality-policy-1345681372276.html. Accessed 13 Sep 2017.
14. Tanner R. Motivation—applying Maslow's hierarchy of needs theory. 2017. https://managementisajourney.com/motivation-applying-maslows-hierarchy-of-needs-theory/. Accessed 13 Sep 2017.
15. US Food and Drug Administration: analytical procedures and methods validation for drugs and biologics. 2015. http://www.fda.gov/downloads/drugs/guidancecomplianceregulatoryinformation/guidances/ucm386366.pdf. Accessed 13 Sep 2017.
16. Annesley TM. Who, what, when, where, how, and why: the ingredients in the recipe for a successful methods section. Clin Chem. 2010;56(6):897–901.
17. US Food and Drug Administration: guidance for industry-good laboratory practices-questions and answers. 2007. http://www.fda.gov/downloads/ICECI/EnforcementActions/BioresearchMonitoring/UCM133748.pdf. Accessed 13 Sep 2017.
18. US Food and Drug Administration: 1981 questions & answers-good laboratory practice regulations. 2007. http://www.fda.gov/ICECI/Inspections/NonclinicalLaboratoriesInspectedunderGoodLaboratoryPractices/ucm072738.htm. Accessed 13 Sep 2017.
19. OECD: OECD series on principles of good laboratory practice (GLP) and compliance monitoring. 1995–2006. http://www.oecd.org/chemicalsafety/testing/oecdseriesonprinciplesofgoodlaboratorypracticeglpandcompliancemonitoring.htm. Accessed 13 Sep 2017.
20. Dance A. Authorship: who's on first? Nature. 2012;489:591–3. https://doi.org/10.1038/nj7417-591a.
21. WHO: laboratory biosafety manual. 3rd ed. 2004. http://www.who.int/csr/resources/publications/biosafety/Biosafety7.pdf?ua=1. Accessed 18 Sep 2017.
22. UCL: UCL code of conduct for research. 2013. http://www.ucl.ac.uk/srs/governance-and-committees/resgov/code-of-conduct-research. Accessed 13 Sep 2017.
23. Cambridge University: research integrity. 2015. http://www.research-integrity.admin.cam.ac.uk/sites/www.research-integrity.admin.cam.ac.uk/files/integrity_leaflet_1115.pdf. Accessed 13 Sep 2017.
24. Freedman LP, et al. The economics of reproducibility in preclinical research. PLoS Biol. 2015;13(6):e1002165. https://doi.org/10.1371/journal.pbio.1002165.
25. Bongiovanni A, et al. Applying quality and project management methodologies in biomedical research laboratories: a public research network's case study. Accred Qual Assur. 2015;20(3):203–13. https://doi.org/10.1007/s00769-015-1132-5.

Quality Tools for the Scientific Research

5

5.1 Tools for Project and Process Quality

From the organisation's point of view, standardised methods can enrich the research environment, making it more beneficial to the studies, freeing up, at the same time, the resources involved in the management in favour of actual research activities. Standardised methods can also provide technical and organisational support to development, testing and technology transfer of the research products.

Quality tools have long been put into practice in the industry. Their translation to the research field, taking advantage of the many years of experience, has already demonstrated the potential to enhance data elaboration and risk assessment, achieving a holistic, integrated management of the project. Table 5.1 shows some tools and methodologies, which can benefit a research study, in terms of both project and

Table 5.1 Tools for project and process quality

Project quality		Process quality	
Planning	• Gantt • PERT • Flow chart	Process control	• Flow chart • SPC • 6 sigma • Lean management
Verification and validation	• DoE • 7 TQM tools	Risk analysis	• P-FMEA
Data analysis	• DoE • 6 sigma	Troubleshooting	• Problem solving
Risk analysis	• D-FMEA • P-FMEA		
Troubleshooting	• Problem solving		

Some quality tools and methods are suggested for supporting the management of project quality in planning, verification and validation, data analysis, risk analysis and troubleshooting, and for the management of process quality in process control, risk analysis and troubleshooting

© Springer International Publishing AG, part of Springer Nature 2018
A. Lanati, *Quality Management in Scientific Research*,
https://doi.org/10.1007/978-3-319-76750-5_5

process. In the next chapter, both project management methodology, such as Gantt and PERT, and methods, such as DoE, FMEA, lean management and 6-sigma, will be addressed.

Moreover, team work, team building, leadership, communication and decision-making are all supported and enriched by skills and techniques that can help a researcher deal with day-to-day laboratory operation with respect to non-scientific issues, such as the management of people or working groups, internal and/or external contacts, execution of the project and management of the working environment. These techniques are addressed in the second part of this chapter.

5.2   7 Basic Tools of TQM

TQM—total quality management—includes the use of simple statistical techniques for managing both productive and organisational processes. The 7 basic tools of TQM, as first outlined by Kaoru Ishikawa, a professor of engineering at Tokyo University, are easy to understand and use, so that they can be utilised by staff at all levels: operators, technical managers and executives at all strategic level. A scientific researcher can use these tools at hand, to streamline the approach to their projects, processes and problems.

- Check sheet: a structured form for collecting and analysing data.
- Histogram: for a set of data, a graph showing how frequency is distributed or how often each different value occurs.
- Pareto chart: a bar graph showing which factors are more significant to the variability of a process.
- Cause-and-effect diagram (also called Ishikawa or fishbone chart): identifies many possible causes for an effect or a problem, arranged into useful categories.
- Scatter diagram: graphs pairs of numerical data, one variable on each axis, to look for a relationship.
- Stratification: separates data gathered from a variety of sources so that patterns can be seen (some lists replace *stratification* with *flow chart* or *run chart*).
- Control charts: graphs used to study how a process changes over time.

The *check sheet* can be adapted to fit a variety of situations, projects and studies. As the requirements for a check sheet are myriad, no pro forma/general scheme can be suggested. Any made-to-purpose check sheet, however, should contain information about data collection (what, where, by which tools/instrument, by whom ...) and it should be designed with a view to later analysis; therefore it should record the essential characteristics (e.g. classification or grouping of data) that will allow further elaboration of data. *Histogram* is the most commonly used graph, so basic and universal that they do not require further explanation in this context. *Scatter diagram* and *stratification* are well known from the study of basic statistics. The less famous *Pareto chart*, *cause-and-effect diagram* and *control charts* are worth an

in-depth analysis. For examples of their application in research, refer to the paragraph dedicated to problem solving, which makes a large use of these tools.

5.2.1 Pareto Chart

The Pareto[1] chart is used to separate the *vital few* from the *trivial many* among the
different factors causing an effect, in other words to identify the causes that significantly affect the outcome—either the process or the product. The Pareto chart is
widely used in all phases of problem solving, in combination with other tools, and
in the management of the process. Starting from a histogram of major effects, categories of identified causes are ordered from those at highest to those at lowest
frequency. The principle of Pareto or *principle of 20–80%* has universal validity and
states that in most phenomena a minimum (20% of the total) set of variables contributes to the majority (80%) of the overall effect. Therefore, acting on and eliminating the appropriate 20% of causes, negative results should decrease by
approximately 80%. For example, to improve by 80% the success of the very defective experiment whose Pareto chart is shown in Fig. 5.1, simply find solutions for
the first two problems by relevance.

When to use a Pareto chart? For example, analysing the frequency of defects in
a process, looking at causes in a process or figuring out what is the most significant

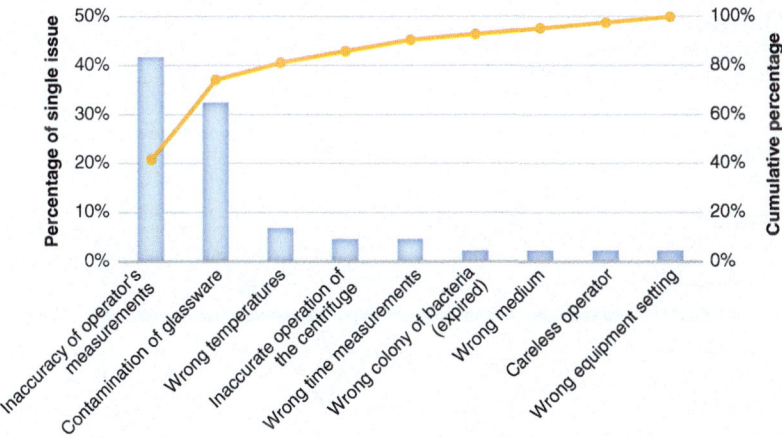

Fig. 5.1 Pareto chart of causes of failure in a research laboratory. The histogram shows the percentage of each cause relative to the total failures in laboratory experiments (left *Y*-axis). The line
is the cumulative percentage read on the right *Y*-axis. Solving and eliminating the first two failure
causes—which together account for approximately 20% of the nine causes—lead to an improvement of 80% on the total number of *failures*

---

[1] Vilfredo Pareto, Italian economist of the early twentieth century, proved that 80% of the world's
wealth is in the hands of 20% of the population.

problem in a process. Care is to be given when defining the categories of factors and/or their limits, but the tool is both simple and powerful.

5.2.2 Ishikawa (Cause–Effect) Diagram

The Ishikawa diagram (named after the scholar who conceived it) is also known as the cause-and-effect diagram or the fishbone diagram. It is most frequently used either to list the possible causes of a problem in any process, organising them into causal chains, or for a qualitative analysis, to determine the characteristics of products and services as targeted on the desired effect.

To correctly use this tool, begin with a clear definition of the effect at issue; this is then inserted into a rectangle connected to a so-called *causal line*. From this line start the primary branches, each one representing a class of potential primary causes. Secondary causes that affect the primary causes can be identified and linked to the primary causes; primary branches can then be broken down into secondary branches. Similarly, tertiary causes may be identified influencing, in turn, secondary causes. The final diagram looks like a fish bone, as in Fig. 5.2. The categories or classes of primary causes may be defined according to the type of problem at issue or to four standard types (in English, 4 Ms):

- Manpower
- Materials
- Methods
- Machines

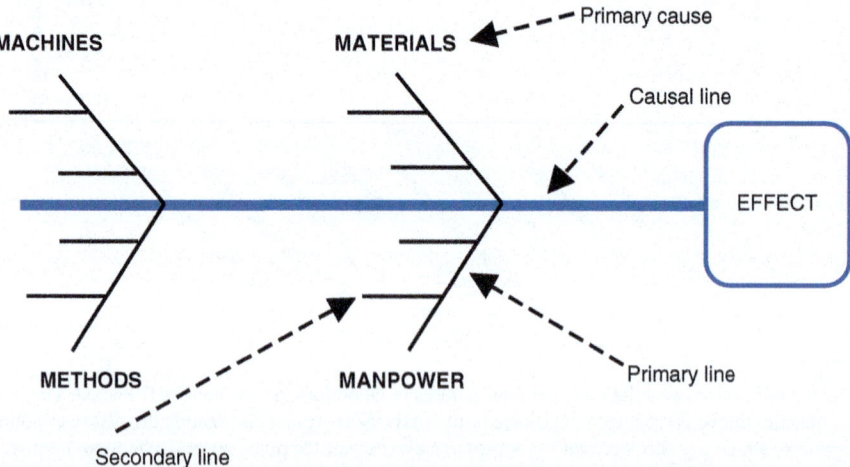

Fig. 5.2 Fishbone or Ishikawa diagram. Starting from the *effect*, a *causal line* is drawn that branches out into four *primary lines*, each of them is given one of the 4 Ms: machines, materials, methods and manpower. Each primary line, if necessary, may branch out in several *secondary lines* or further higher order lines

A fifth branch can be added for the *environment*, called by the French word *Milieu* as a fifth M. More recently, a sixth M for management has been proposed. Others use 4P, standing for parts (raw materials), procedures, plant (equipment) and people.

The result of the first analysis is then *reduced* by evaluating each cause to determine whether it is applicable and whether realistically it is related to the effect.

When completed, an Ishikawa diagram can look rather complicated: even a simple diagram consisting of 4 categories, each with 4 primary lines and possibly 4 secondary lines, could identify 64 causes to be investigated. Conversely, too simple a diagram may denote a scarce knowledge of the process under investigation, as K. Ishikawa himself originally pointed out. In that case, the diagram may highlight areas of the process worthy of deeper understanding.

Overall, the fishbone diagram is easy to understand and use, especially when the context is well known and understood. It has the disadvantage of ignoring the interaction between the causes and the risk of being tainted by personal opinion, so it is recommended to use it in teamwork and in conjunction with other investigation and verification tools.

5.2.3   Control Charts and SPC

The control chart was developed by W.A. Shewhart in 1924 for the Bell Telephone Laboratories. It is designed to monitor the variability of a process over time and is the most common tool used in the statistical process control (SPC). SPC is a statistical method that allows containing the variability of a process output within specific limits; this guarantees the compliance of the output to quality and performance requirements. The control chart indicates whether the process is under control and how to address appropriate corrective actions, against either *common (or random) causes* or *special causes* of variability. The former are due to the structure of the system and are not removable without a significant change to the process, while special causes are due to *accidents* and are identified analysing the drift of the process parameters. For example, in the first category can be included the temperature that may cause drifts on a measure made with a specific instrument. If more precision is needed, then either modification to the instrument (e.g. thermally isolating the sensing device, modifying the original device) or use of different equipment (i.e. changing the measurement system) will deliver more precise, temperature-independent measures. The second category, *special causes,* includes failures of the instrument, wear and tear of the sensing element or misuse. In these cases, the problem can be solved with the equipment maintenance or the user's training.

There are two classes of control charts:

- Variable control chart—used to keep under control a measurable characteristic: size, weight, concentration, etc.
- Attribute control chart—used when a characteristic is to be classified as compliant or non-compliant and the judgment is only qualitative

The *Xbar-R chart* is used to analyse the process parameters or characteristics of the product. Once the characteristic to be measured (one for each control card) and the tools to be used are identified, samples are taken at fixed times, each sample consisting of a defined number of observations. Average and range (difference between minimum and maximum values of the sample) of each sample are recorded on an XY-graph with time on the abscissa and the values on the ordinate. On the ordinate are shown the lower and upper limits allowed (tolerances). The variability of the samples as described by the tolerance of mean and range is typical of the process, is due to random causes and is determined at the design stage. A value outside the tolerance limits means that the process is out of control due to a special cause. Analysing the temporal sequence of samples, information can be drawn to identify the cause of process deviations from the expected variability. The control chart can also be used for improvement purposes. Analysing the samples falling near the limits and trying to identify causes of larger variability will inform/guide related corrective actions. Figure 5.3a shows an example of chart Xbar-R, with the average X, the range R and the upper/lower limits for X and upper only for R.

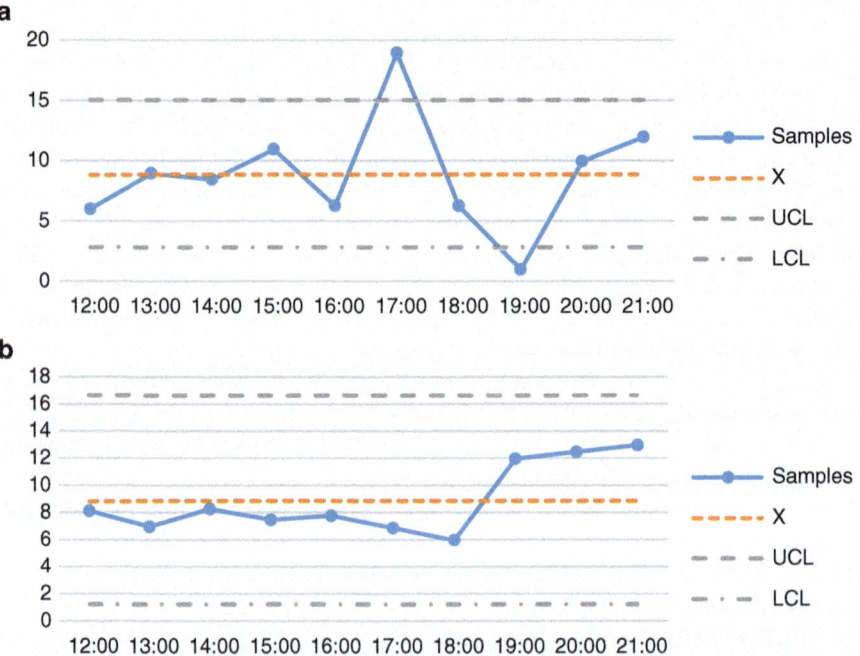

Fig. 5.3 Two examples of X bar control charts. The measurements from the samples are plotted each hour. In (**a**) the trend shows an out-of-control datum reaching over the upper limit at the sixth point (17:00 h) and another one below the lower limit at 19:00 h (eighth point). In (**b**) seven consecutive points below the average value, although still within the limits, are indicators of an alarm condition

The *PN chart* and the *P chart* are used to control the number of defective units, and then to monitor samples by attributes: the P chart takes into account the data percentage (e.g. defectiveness), while in the *PN chart* the value is normalised with respect to a predetermined quantity. The U chart and the C chart are used to control the defects on the products: the U chart measures the number of defects per unit of measurement (e.g. per length, per area), and the second the number of defects in a *constant sample* (e.g. in a single device).

In designing a control chart, care is to be taken first with the choice of chart, depending on the kind of process and control envisaged. The definition of control values can be made on a historical basis or on a statistical basis. When choosing the historical basis one must have a thorough knowledge of the process and its parameters. The statistical basis entails the calculation of—for example—±3% of the typical value established by design. Control values can also be set according to externally imposed targets, for example ±10% of the assigned budget. The dimension of the sample and the sample frequency can be determined upon knowledge of the process and some considerations of cost/benefits: small samples taken frequently are preferred in large productions, where several types of variability factors can be encountered. In other situations, large samples taken less frequently may be the most efficient choice.

Figure 5.3 illustrates how the process can be controlled and even non-conformities can be anticipated with the use of control charts. In the first chart the trend of samples (plotted each hour) shows an out-of-control datum reaching over the upper limit at the sixth point (17:00 h) and another one below the lower limit at 19:00 h (eighth point). This situation must be addressed with corrective actions. The second chart shows a trend of seven consecutive points below the average value. This condition is considered hazardous for the process quality and may be addressed with a preventive action. The value of seven points all below or above the average is empirically defined.

5.2.4   Flow Chart

Even if not comprised among the seven tools of quality, flowcharting has a relevant position among quality tools, because it allows describing a process in a simple and effective way, at the same time calibrating the level of detail.

Flowcharting is born in the informatics world and very soon it migrated everywhere a process must be described. The start and the end of the flow are represented with elliptical symbols. Each activity of the process is set in a box. Each box is connected to the previous and the next one by a link. When the activity entails a choice among different alternatives, a rhombus substitutes the box. When the activity is complex and deserves a more detailed description, it can be considered as a subprocess (*predefined process*) and represented with a box with doubled sides. Figure 5.4 shows most used symbols in flowcharting.

A flow chart may be as simple as the one showed in the first column of Fig. 5.5. The same figure illustrates how a flow chart can be used to plan a process, indicating who will perform the tasks (delegate), when they have to be completed, which are

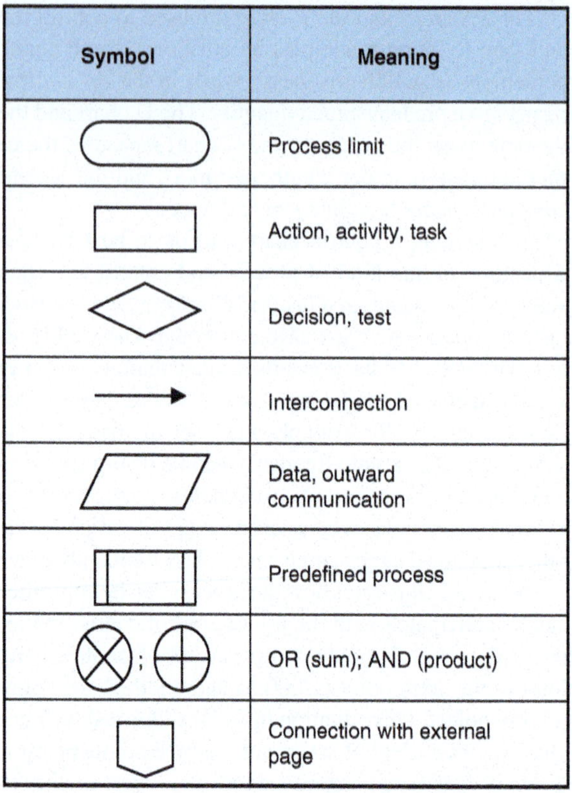

Fig. 5.4 Symbols of flowcharting. The first four are the most used: Process limit, *action*, *decision* and *interconnection*

Symbol	Meaning
⬭	Process limit
▭	Action, activity, task
◇	Decision, test
→	Interconnection
▱	Data, outward communication
⬛	Predefined process
⊗ ; ⊕	OR (sum); AND (product)
⬠	Connection with external page

the documented references and what will be the output of tasks. The same form can be employed to sketch a procedure, substituting the column *Reference* with a short description on *How* to do the task.

Actually this scheme is a simplification of a more formal one, named SIPOC from the acronym of Supplier, Input, Process, Output, Customer that are the titles of the five columns. In this case, the flow chart can be inserted in the third column, *Process*. The SIPOC diagram is employed in the six sigma approach[2] for analysing the process to be improved. Figure 5.6 shows another graphical representation of the SIPOC diagram.

Another most used scheme for flowcharting is structured as a matrix (Fig. 5.7) and often called *cross-functional flow chart*, where activities are split according to the responsibility, in order to make clearer the flow through the various functions.

Flowcharting is most useful when working as a team on a process to be designed or improved: it helps not only in making a clear scheme of the flow of activities but also of the data to be collected, the materials required and the outputs expected, thus facilitating for team members the sharing of information and the creation of solutions.

---

[2] See Chap. 6—Lean Management and Six Sigma.

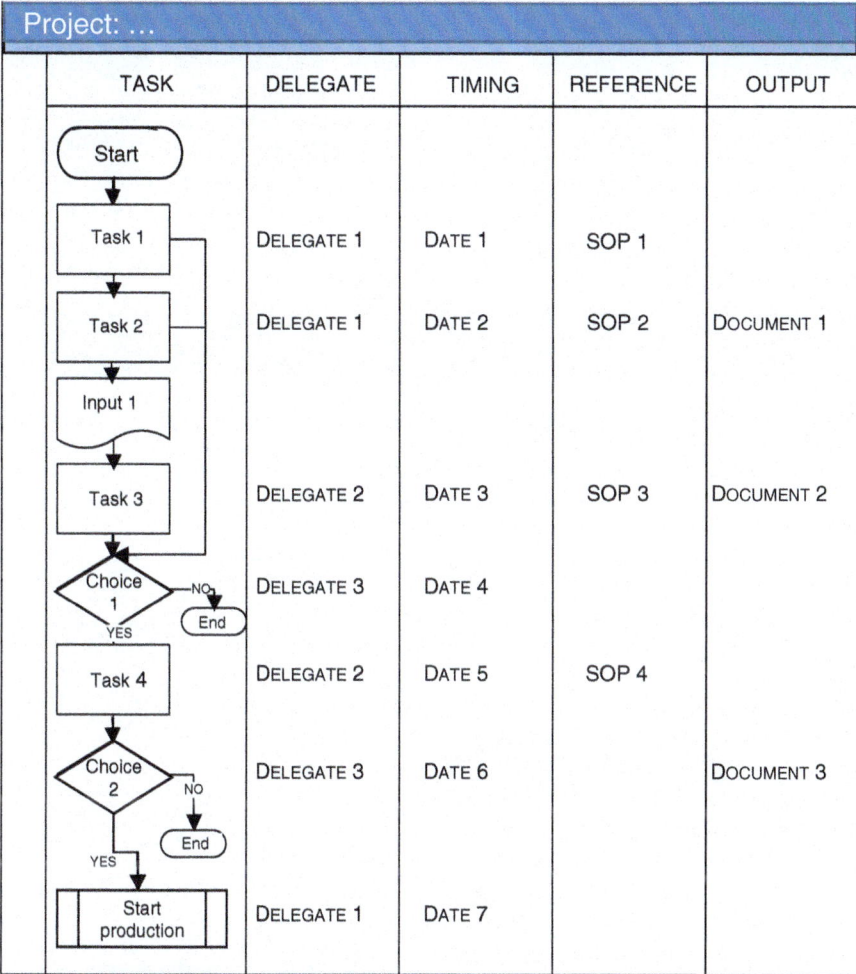

	TASK	DELEGATE	TIMING	REFERENCE	OUTPUT
	Start				
	Task 1	DELEGATE 1	DATE 1	SOP 1	
	Task 2	DELEGATE 1	DATE 2	SOP 2	DOCUMENT 1
	Input 1				
	Task 3	DELEGATE 2	DATE 3	SOP 3	DOCUMENT 2
	Choice 1 / End	DELEGATE 3	DATE 4		
	Task 4	DELEGATE 2	DATE 5	SOP 4	
	Choice 2 / End	DELEGATE 3	DATE 6		DOCUMENT 3
	Start production	DELEGATE 1	DATE 7		

Fig. 5.5 Planning with a flow chart. The flow chart describing tasks and decisions is drawn in the first column of a table. The other columns are for *delegates* (the person charged for the task), *timing* (the schedule), *reference* (any prescriptive document inherent to the specific task) and *output* (any kind of outcome expected from the task)

5.2.5 Example of Application: Protocol Flowcharting and Process Control

An innovative optical platform for ion channel drug screening (ionChannelΩ), based on a proprietary approach, was developed by a multidisciplinary team in the Alembic laboratory of the San Raffaele Scientific Institute. The system is complex, including a microscope, a camera, a laser, a liquid dispenser and various devices for precision motions and positioning. Briefly, cells expressing the channel of interest are loaded with a fluorescent voltage-sensitive dye and the effect of a drug is

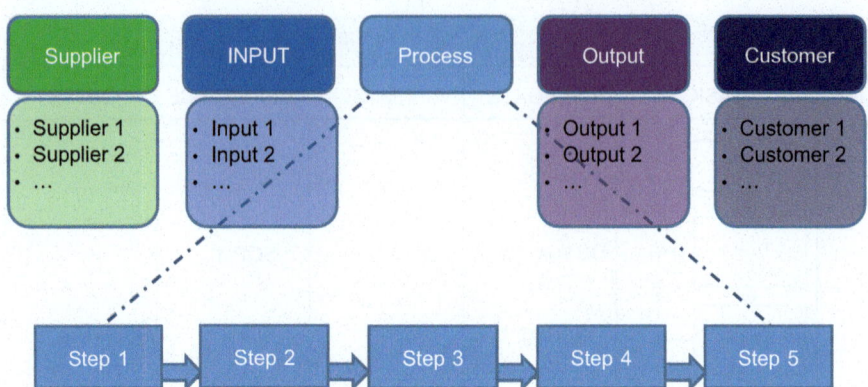

Fig. 5.6 The SIPOC diagram. The diagram is structured in five columns. In the first one, suppliers, namely who are expected to provide the inputs, are listed. The second column is dedicated to the inputs needed by the process. In the third column the process is deployed into its main activities. Columns 4 and 5, respectively, refer to the expected outputs and their recipients

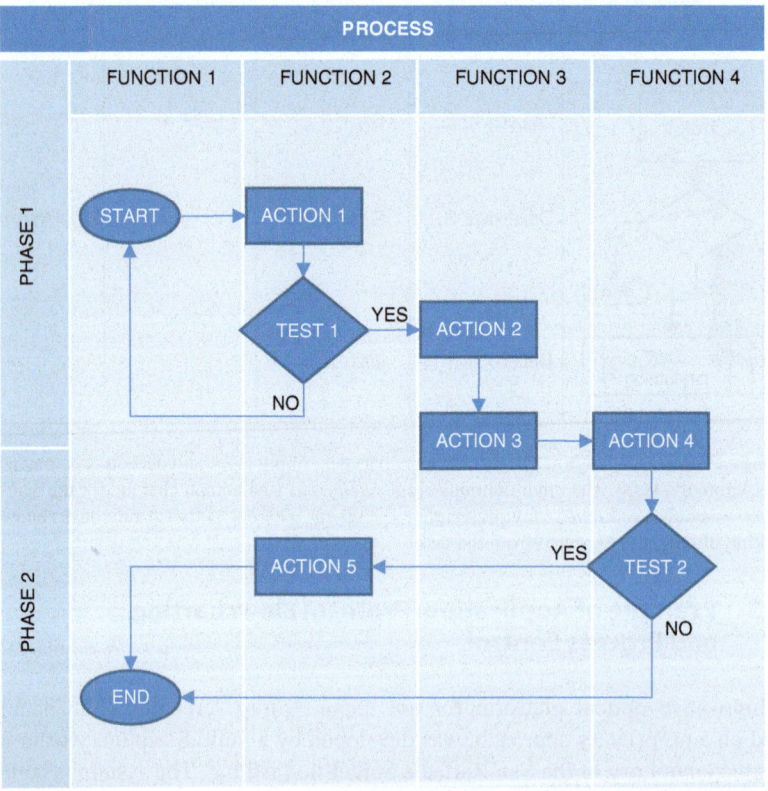

Fig. 5.7 Cross-functional flow chart. Each column represents a person or function charged with one of the tasks. Each task is put in the column of the person/function in charge of its execution. The process may be subdivided into phases, using the first column to identify them

revealed by the fluorescence values recorded before and during exposure to electrical stimulation.[3] Many quality tools were used throughout the project; among them flowcharting was employed to design the protocols of stimulation and analysis. Figure 5.8 illustrates the high-level flow chart; for the sake of simplicity, only the process steps for one lane analysis are reported. Cells expressing the receptor/channel of interest and the drugs under evaluation are prepared; then the multiwell tile containing the cells is positioned into the equipment and the objective lens is manually focused, based on the sharpness of cell images. At this point, the equipment is ready to start the acquisition of images before and after electrical stimulation, both in the absence and in the presence of the test compound. This screening protocol is applied to multiple sampling areas within each single well and then to all the wells of the lane, in sequence. In order to do this, the multiwell is automatically controlled to change the sampling area, the focus is automatically set and the cells are electrically stimulated: within this sequence of events, relevant images are captured at precise time points. At the end of the lane, tools for electrical stimulation and drug dispensing are automatically washed. The equipment is then ready for the analysis of the next lane, while data from the previous one can be elaborated to draw a concentration–response (C–R) curve.

Drawing the flow chart for this protocol streamlined its definition, allowing the identification of each single step and the correct sequencing of the entire process, and defining the times and characteristics for all operations. The flow chart also allowed optimal placement of controls to guarantee correctness and repeatability of the process. In Fig. 5.8, the first column in the table on the right of the flow chart describes the controls considered necessary by analysing the flow chart. For example, to assure standardisation of the test vehicle, it was deemed necessary to write a document for the supplier specifying the characteristics of the cell line to be used, the controls to be executed and the conditions of stock and delivery. In the same way, the reproducibility of dispensing volumes was checked by weight with an analytic balance and the correct dilution sequence of the drug was tested with a colorimeter in order to verify the correctness of the C–R curves.

The controls identified through the analysis of the flow chart were applied and found to be appropriate and useful. Then another quality tool, the failure mode and effect analysis,[4] was used to investigate the risk of errors in the protocol. Again, the input for the FMEA is a flow chart that allows identifying and analysing, in turn, each process step. The FMEA application demonstrated the need for further controls, listed in the second column of the table in Fig. 5.8. As an example, it highlighted the need to further train the operator providing them with a specific checklist for more delicate operations. It pointed out the importance of periodic checks, calibration and maintenance of devices, and specific controls such as the one listed as *Xbar-R chart for electrode parameters*. This last control deserves a further explanation: the electrodes used to give the electrical stimulation to the cells can accumulate organic residue that endangers the measurements, modifying the readout of the voltage, given a fixed

---

[3] See EP2457088/WO2011009825 patent for more details.

[4] See Chap. 6—Failure Mode and Effect Analysis.

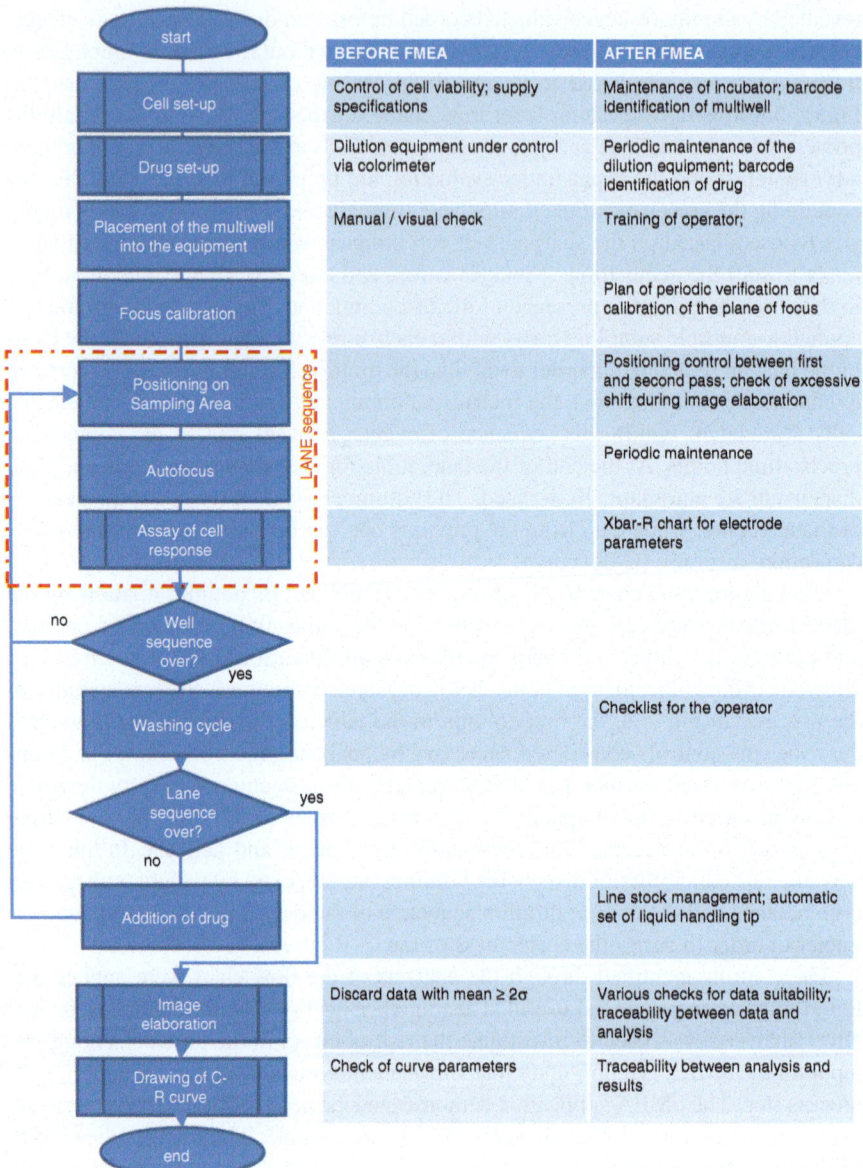

Fig. 5.8 Flow chart of the drug screening process (OMICS project). The first steps of the flow chart concern the preparation of drug and test vehicle. After the calibration of focus, the *lane sequence* is started, comprising *positioning*, *autofocus* and *assay* of cells, and repeated for every well containing cells. After the first lane sequence and the washing cycle, the drug is added and a second lane sequence is executed. Finally, data are elaborated to draw a C–R curve. The column *Before FMEA* describes the controls considered necessary by analysing the flow chart. The column *After FMEA* shows further controls indicated by the FMEA application

current. It was decided to measure the ratio between the voltage and the current, namely the resistance between the electrodes, and to plot those values in an X–R chart. The form of the X–R chart was slightly different from the usual: one chart for each lane was generated, composed by 12 values: 9 for the wells with cells and 3 for the wells devoted to washing. In this chart, the X is the mean of the values of each intra-well sample, while the R is the difference between the lowest and the highest of those values. Figure 5.9 shows two examples of this kind of X–R chart, where the mean X is shown in the left axis and the range R is shown in the right axis. Diagram

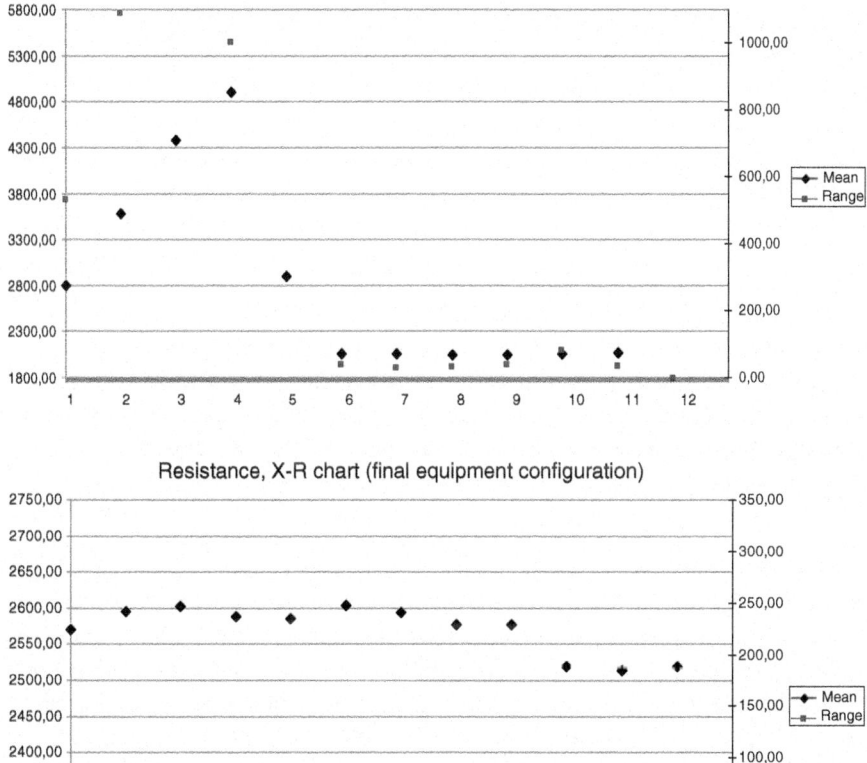

Fig. 5.9 X–R chart for the resistance seen from the stimulating electrodes. Each sample corresponds to a well. The points for *mean* are the mean of several readings of the resistance between electrodes in each well. The points for *range* are the difference between the lowest and the highest values recorded in the well. (**a**) During equipment development, values for mean and range were high (more than 4000 Ohms for mean and 1000 Ohms for range) and not stable. (**b**) In the final equipment configuration, mean was stable at approximately 2550 Ohms and range not over 50 Ohms

a) shows X–R chart during the development and tuning of the equipment; the process was not under control and both X and R were unacceptably large and variable. Diagram b) shows that in the final configuration of the equipment, the variability of both X and R was acceptable and the reproducibility from well to well is very high. The upper (for X and R) and lower (for X only) limits were statistically determined at ±3σ around the typical value. It was also planned to generate an automatic alarm for value readings outside the allowed range.

5.3   Decision-Making

Not making a decision is still a decision. This aphorism is so popular that it is very hard to find its actual origin. Someone attributes it to an ancient English proverb, and others to J.W. Goethe. The verb *decide* comes from the Latin *de-caedere*, which means *to sever, to cut*. Decide is equivalent to discard—cut—all possible alternatives to achieve the goal. The objective of a decision may be described as maximising the benefits, that is, the intrinsic worth of the decision-maker's ultimate decision. A good decision minimises the emotional burden due to the presence of conflicting values between alternatives [1], as well it minimises the cognitive effort necessary to acquire and process information.

Good decisions are based on a clear analysis of the situations and the alternatives, and cannot be made without gathering information and data. The best way to make a decision is to follow a rigorous method. It must be distinguished between a *wrong* decision and a *bad* decision. A wrong decision is choosing to invest in stock, after a careful evaluation of alternatives just a few days before the outbreak of a global financial crisis. It ends in a mistake, but the result does depend on the method followed, which could not take into consideration all aspects of the financial world and the associated risks. A bad decision is often done intentionally without a proper evaluation, such as choosing to invest in equity just on the basis of rumours and hearsay. The wrong decision is due to the imponderable, while the bad decision is due to a lack of a correct decision process. This distinction separates the results of a decision process from the process itself: in the first case, the process has been pursued but chance has played against a good outcome; in the second case no defined decision process has been followed, ruling out any possibility of arriving at a good result. The difference is that chance cannot be controlled, but the process can, and this can help in reducing the role of chance. It is worth noting that considering a mistake as a wrong decision relieves the individual scientist or team from all responsibility.

The process suggested to arrive at a good decision involves several steps, which sound very similar to the well-known PDCA cycle:

1. Define the goal
2. Collect information
3. Generate achievable options
4. Make the decision, evaluating risks and assessing consequences
5. Implement and evaluate

The model for decision-making by the Nobel Prize laureate Herbert Simon is well known in the field of management and separates the process of decision-making into three main steps: intelligence, design and choice. In the intelligence phase, data are collected and examined and the problem is properly defined. In the design phase, alternatives are generated, evaluated and compared, with respect to risks, costs, probability of success and other criteria linked to the nature of the decision to be made. In the third phase, choice, one of the alternatives is selected, based on the selection criteria. If the solution is not satisfactory, the process must be started again from the intelligence phase, and the collection and analysis of more data.

Decision-making can be considered as a stand-alone discipline or—even better—a phase of the more complex process of problem solving. In the section below, dedicated to problem-solving techniques, are illustrated some tools that can be used to streamline the analysis and guarantee the related outcomes.

5.4   Problem Solving

Problem solving is a method—i.e. a formalised process—aimed at solving a problem. But what is a problem? And what does it mean to solve it? Seeking the aid of well-known dictionaries[5] leads to these definitions:

Problem: something that is difficult to deal with: something that is a source of trouble, worry, etc.

To solve: to find a way to deal with and end (a problem); to find a solution, explanation or answer for

To maintain a scientific approach, however, a less qualitative definition is needed. Therefore, we can refer to a *real difficult situation* and to a *target ideal situation*, and define the problem as *a deviation between actual data and reference data, which generates negative effects.* Now it is easier to define quantitatively the resolution of a problem: to solve a problem is *to eliminate the causes of the negative gap, so that it will not occur again.* This definition relies upon the core concept of problem solving: solving a problem is identifying actions that prevent the problem from happening again, *and* addressing actions to the causes of the deviation: it is not to limit the effects of the deviation or find actions to remove symptoms.

In order to solve a problem, the most common approach is trial and error, which is neither efficient nor effective and inevitably affects costs, as well as lowers the morale of the people trying to deal with it. A scientific approach to problem solving, i.e. a structured method, offers several advantages over an approach simply based on instinct or experience. The solution, identified after detailed analysis and assessment of alternatives, has a higher degree of certainty or success. Following a decision-making process that evaluates risks reduces the uncertainty of the final decision. Testing only accurately chosen and validated solutions ensures a lower

---

[5] In this case: *Merriam-Webster.com*. Merriam-Webster, n.d. Web. 17 Mar. 2014. http://www.merriam-webster.com/dictionary.

cost of experimentation. Finally, if the solution is conceived by a team working within an organisation, a broad consensus will ensure an easier implementation.

A problem-solving method can also be applied to a project, viewing the goal as the gap between the current and the ideal situation; in this case, the deviation is not a hardship but a positive evolution from the current situation. A research project can be thought of in a similar way, the deviation being the gap between current knowledge and confirmation or refutation of a scientific hypothesis.

Problem solving is a team job, where skills and competences are the core resources, together with a good leadership of the group. Thus, it is important to select the appropriate team members for the contribution that they can make to the process, and to provide them with training on team work. Skills in problem solving are a powerful competence that fosters any team work. In the following sections, team work and related tools will also be discussed.

5.4.1   The Problem-Solving Wheel

The process of solving problems needs both experience and creativity. In order to maintain the correct balance between them, it is useful to follow a formalised method. In the literature, many methods of problem solving can be found, usually graphically represented. In all of them, the process begins with collecting data and analysing them, then searching for root causes, defining one or more corrective actions, implementing the intervention and controlling the outcome with the possibility of a resumption of the process, if the outcome is not satisfactory. For this reason, the process is often represented by a wheel composed of 6–8 steps.

One of the possible methodological approaches is shown in Fig. 5.10 and follows six steps:

- In the step *identify and select problem*, a clear description of the problem as a deviation from the ideal situation is the purpose of the phase.
- The *process of problem analysis* or *diagnostic process* is the focal point of the problem-solving wheel: starting from the analysis of the deviation it identifies one or more underlying causes.
- The *process of generation* of *solutions* identifies most actions intended to remove the root causes of the deviation, each one appropriately described.
- The *decision-making process* evaluates each proposal developed from the previous phase, with respect to constraints and preferences, and analyses the risks associated, on both the impact on the implementation phase and the cost. This leads to the implementation of what has been decided to be the solution to the problem.
- The *process of planning and implementing* generates an action plan to implement the solution chosen in the previous phase: in order to *do it right the first time* it must therefore include checks on the critical points and possible emergency solutions. The result of this step is the implementation of the solution to the problem.

Fig. 5.10 The problem-solving wheel. The cycle is composed of six steps, starting from *identify and select problem* to *monitor and evaluate*. If the result of the last phase is positive and the problem has been solved, the cycle ends. Otherwise it continues, starting again from the first step and taking into account data and information collected in the first run

- In the process *monitoring and evaluating* the result is achieved when it is proved (i.e. measured) that the deviation no longer exists. Also, in this phase, results are recorded and disseminated. Once the actions are concluded, it often happens that this step is neglected: however, it is essential to build a culture and a knowledge base, maintain updated documentation about each issue and nurture a technical memory to be made available for the solution of similar cases. In the end, this approach builds enriched and improved business processes, taking inspiration from the experience and the solutions identified.

Table 5.2 An example of problem solving as applied to *the choice of a course of studies after secondary schooling*

Problem	Choosing a course of study after secondary schooling
Problem identification	Registration of personal inclinations and constraints (economics, logistics, social aspects …) Collection of information about universities Collection of information on the professional opportunities
Analysis of the problem	Correlation between commitment to study, professional career and personal expectations Correlation between economic burden and professional opportunities
Generation of solutions	List of degree programmes that meet the requirements Collection and organisation of related information
Decision-making	Assessment of risks and opportunities for individual options Assignment of weights to options Choice of degree course and possible universities
Plan and implement	Enrolling and carrying out selection test if required Checking final selection Enrolling at university Beginning to study
Monitor and evaluate	Checking (*in itinere*, exam results) study results Assessing personal satisfaction Possible corrective action (change degree course or domicile, etc.)

The process for solving the problem is organised in six steps according to the problem-solving wheel. Main activities are listed for each step

Table 5.2 shows an example of the application of problem solving in a practical situation of daily life.

5.4.2   The Tools of Problem Solving

Table 5.3 shows the most common troubleshooting techniques for each problem-solving step. A few of them will be briefly detailed in the following, while the others are described in different sections of this text. For an in-depth discussion of the tools not described in this book, refer to [2] or [3].

Current Reality Tree. The current reality tree is one of the *thinking processes* developed under the theory of constraints (TOC) by E. Goldratt [4]. It is mainly used when dealing with an organisation and aimed at representing most organisational problems at the same time. It can be used in first phases of the problem-solving process, to describe the situation, collect data and seek the main structural causes.

A team of experts or persons in charge makes the drafting of the current reality tree. Undesirable effects (UDE) are considered symptoms of perceived or hidden root causes and are represented in a simple cause–effect network, filled and completed through team work. The graphical representation facilitates verbalising symptoms and causes—the more clearly and unequivocally described, the better the current tree works—helping to identify connections and dependencies between them, as well as the preconditions or neutral effect, which favour the emergence of negative effects.

Table 5.3 The tools of problem solving

Step	Tools/techniques	Reference
Problem identification	Gap analysis	
	Brainstorming	p. 137
	Benchmarking	
	Surveys and interviews	
Analysis of the problem	Data collection sheet	
	Flow chart	p. 111
	5 W + 1 H	p. 123
	5 Whys	p. 125
	Pareto chart	p. 107
	Stratification analysis	
	Current reality tree	p. 122
Generation of solutions	Control charts	p. 109
	Correlation chart	
	Mind mapping	
	Brainstorming	p. 137
	Ishikawa diagram	p. 108
	Flow chart	p. 111
	5 Whys	p. 125
Decision-making	Force field analysis	
	Constraints window	p. 126
	Decision grid	p. 127
Plan and implement	Flow chart	p. 111
	Gantt	p. 151
	PERT	p. 149
Monitor and evaluate	Gap analysis	
	Final report	p. 132
	Technical memory	p. 132

For each step of the problem-solving wheel, most used tools are suggested and (where appropriate) accompanied by a reference to the page of this text where they are discussed

It is strongly suggested to seek out the often hidden negative feedbacks which, however, are easier to discover by using the graphical representation. As a team tool, the current reality tree is also intended to achieve substantial convergence of opinions on the shared description of the situation. The chart then provides quantification of the influence of the main causes on the symptoms, thus allowing prioritisation of the needed interventions. Indicators may be associated with the symptoms to monitor the effectiveness of actions undertaken during the implementing phase.

Figure 5.11 shows the scheme of a current reality tree.

Kipling's Servants (5 W + 1 H). One of the simplest tools assisting problem solving is the so-called *5 W's + 1 H*. It seems to have a noble origin, since Rudyard Kipling in his short poem *The Elephant's Child* suggested to employ six zealous servants:

I keep six honest serving-men
(They taught me all I knew);
Their names are What and Why and When
And How and Where and Who.

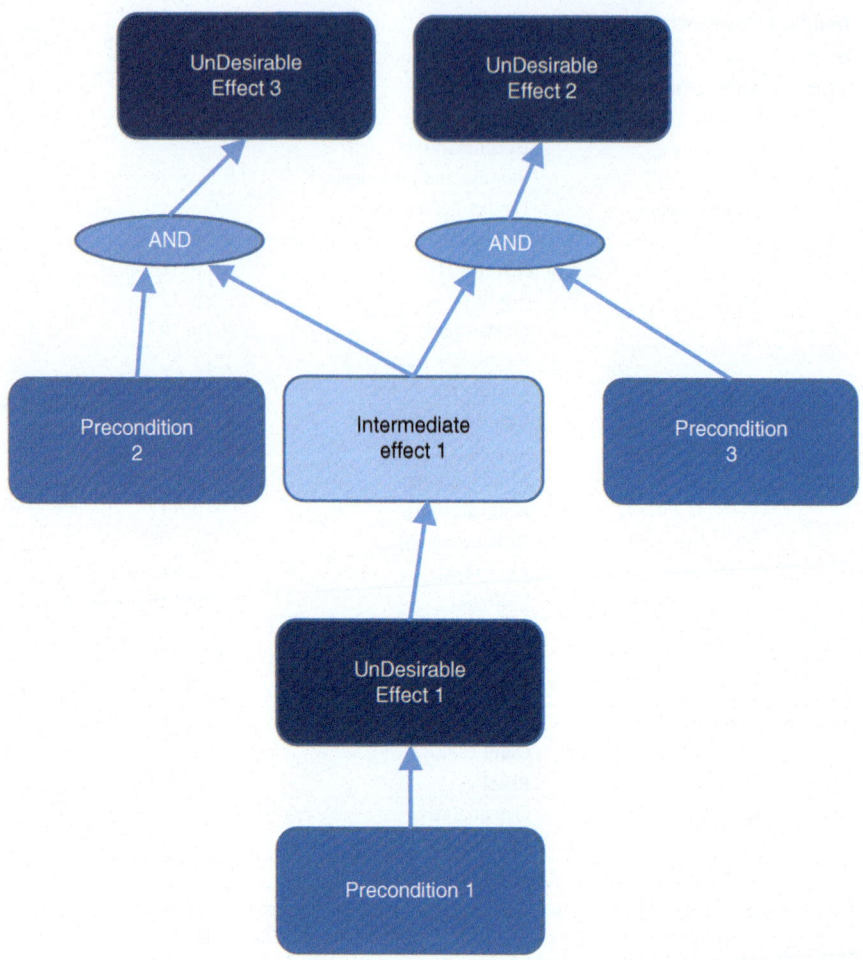

Fig. 5.11 Scheme of a current reality tree. The scheme is drawn starting from the undesirable effects (UDE), identifying the preconditions that lead to the UDE and linking them all in a cause–effect diagram that may highlight intermediate effects and more UDE

This clever idea brought to the development of a list of questions (a checklist), widely used in journalism and investigations. Some variation of the original method may be found: some add a second H for *How much* leading to *5 W + 2 H*, and others substitute the original H for *How* with *How much* and add a sixth W standing for *Which way* (6 W + 1 H).

The order of the questions is not important. This simple method can be used to describe the initial situation, identify causes and generate solutions for improvement.

The already cited paper by T. M. Annesley [5] reports an interesting example of the application of this technique, although outside the field of problem solving. The author suggests the use of the *Kipling Servants* when writing a scientific paper, to ensure the description and inclusion of all information required for proper discussion and for the reader's comprehension.

5 Whys. The *5 Whys* is an iterative questioning technique used to explore the cause-and-effect relationships underlying a problem. The 5 Whys can be used independently or as part of a cause-and-effect diagram. The problem solver proceeds by formulating the question *why?* five times in five progressively deeper degrees of detail. This approach has been devised by Taiichi Ohno, the father of the Toyota Production System also known as lean production or lean management.[6] To him is also owed this significant and widely known example of the application of the 5 Whys:

1. *Why did the robot stop?* The circuit has overloaded, causing a fuse to blow.
2. *Why is the circuit overloaded?* There was insufficient lubrication on the bearings, so they locked up.
3. *Why was there insufficient lubrication on the bearings?* The oil pump on the robot is not circulating sufficient oil.
4. *Why is the pump not circulating sufficient oil?* The pump intake is clogged with metal shavings.
5. *Why is the intake clogged with metal shavings?* Because there is no filter on the pump.

This simple example clearly shows how to get to the root cause of the problems, avoiding stopping the investigation at the first symptom. Its application, however, may be not as easy as it seems at first sight. The method can also be applied graphically, generating a tree of cause and effects. It is important not to stop enquiring, until a *structural* cause has been identified.

Interrelationship Diagram. Developed by the Society of Quality Control Technique Development in association with the Union of Japanese Scientists and Engineers (JUSE) in 1976, the interrelationship diagram (ID) helps to identify the intertwined relationships in a complex problem. *The intent of the ID is to encourage practitioners to think in multiple directions rather than linearly so that critical issues can emerge naturally rather than follow personal agendas* [6].[7] The diagram is built during a brainstorming session, starting from the main problems or causes (items) that emerged through previous analysis (e.g. Ishikawa) and different sources, highlighting mutual relationships. It is useful to ask questions like *Is this item related to any other?* or *Does this item cause or influence any other item?* and then drawing connecting arrows accordingly. The items with several outgoing arrows are likely to be the most important root causes, while the items with several incoming arrows are expected to be the most frequent effects. Some care must be taken in rephrasing concepts and collecting consensus from the working team, including

---

[6] See Chap. 6—Lean Management and Six Sigma.

[7] Reprinted with permission from Quality Management Journal © 2005 ASQ, www.asq.org.

redrawing the diagram till a good representation and a common agreement are reached.

Selecting a Tool for Root Cause Analysis. The most used tools for root cause analysis are the cause-and-effect (Ishikawa) diagram, interrelationship diagram and current reality tree. A. Mark Doggett, Humboldt State University, presented an interesting article [6] about how to choose among them based on appropriate effectiveness criteria, and providing a useful table of comparison. Unexpectedly, the conclusion is that the main issue is the team, who must choose any one of the tools, as good results cannot be achieved without them. Moreover, the effectiveness of the problem-solving process rests on a well-trained team, both for the technical aspects (as root cause analysis tools) and for team dynamics.

Constraints and Preferences. Comparing the possible solutions against the characteristics and limitations of the system is preferable to comparing the possible solutions against each other. The identification of *constraints* and *preferences* helps in discerning between the options allowing ranking with respect to the goal envisaged, thus helping to determine the final choice:

- *Constraints* are conditions that must be strictly taken into account by the solutions on pain of exclusion.
- *Preferences* are criteria for qualitative assessment helping to reach a final ranking.

It is helpful to use a *Table of Constraints* or *Constraints Window* (Fig. 5.12) to manage and summarise results of the first screening. Options compatible with all constraints are kept; options that do not meet even one constraint are marked with *no* and rejected; options that have a dubious compatibility are marked with a *?* to be kept for further analysis.

1. Aim									Results
	Target constraints				Context constraints				
	Constraint 1	Constraint 2	Constraint 3	Constraint 4	Constraint 5	Constraint 6	Constraint 7	Constraint 8	
Solution 1									
Solution 2									
Solution 3									
Solution 4									
Solution 5									
Solution 6									

Fig. 5.12 Constraints window. The aim is described on the top of the matrix. Options are listed in rows (*solutions*), while constraints, possibly grouped into *target constraints* and *context constraints*, are listed in columns. A final column *Results* summarises the compliance of each option to its constraints with a logical AND

Decision Grid

1. Aim of the decision:

2. Criteria ↓	3 Options →	Weight	Option "P"		Option "Q"									
			Assessm.	Result (AxW)	Assessm.	Result (AxW)	Assessm.	Result (AxW)	Assessm.	Result (AxW)	Assessm.	Result (AxW)
criterion Y1		W1	A_{P1}	S_{P1}	A_{Q1}	S_{Q1}						
criterion Y2		W2	A_{P2}	S_{P2}	A_{Q2}	S_{Q2}						
criterion Y3		W3	A_{P3}	S_{P3}	A_{Q3}	S_{Q3}						
criterion Y4		W4	A_{P4}	S_{P4}	A_{Q4}	S_{Q4}						
criterion Y5		W5	A_{P5}	S_{P5}	A_{Q5}	S_{Q5}						
TOTAL				T_P		T_Q		0		0		0

Fig. 5.13 Decision grid. The aim of the decision is described on the top of the matrix. Criteria are listed in rows and solutions (*Option* X) are assigned to columns. Each criterion is given a *weight* (1 = lowest to 5 = highest) based on its bearing on the final decision, and each option is assessed against each criterion (1 = lowest to 5 = highest). For each option, a partial *score* is calculated multiplying their assessment against the criterion and the criterion weight. All results for each option are then added to give its *total* score. Total scores are compared to choose the best option

Preferences are used as criteria for filtering the brainstorming output and may be managed with a *decision matrix* (or *grid*). First, the aim of the decision must be clearly defined. Then the team must identify the criteria that are used to characterise each solution. Figure 5.13 shows an easy-to-use scheme: the options (P, Q) are evaluated (A) for their compliance with the selection criteria (Y) and then weighted according to the importance given to each individual criterion (W). The values thus obtained (S = W × A) are added for each proposal to obtain a total score (T), which permits the ranking of the options.

5.4.3   Example of Application: Improving Scientific Productivity in a Laboratory

A research laboratory trying to improve scientific productivity, after analysing the situation, identified four options to be selected: holding meetings for exchanging competences, training on specific subjects, defining clearer roles and tasks and making working hours more flexible. The criteria for choice were identified as lower risk, better human resource management, better methods, better work climate and organisational effectiveness (Fig. 5.14). Preferences and options were weighted on a scale from 1 to 5, and then the weighted assessments for each option were calculated (grey columns). As an example, the solution *meeting for exchanging competences* was judged very effective for lowering risk and enhancing team work,

1. Aim of decision:
Improve productivity of a research laboratory

2. Criteria ▼ ──────────►	W	3 Options meetings for exchanging competences A	Score	individual training of specific subjects A	Score	roles and tasks better defined A	Score	flexible working hours A	Score
risk	4	5	25	3	15	3	15	2	10
human resource management	5	5	25	5	25	5	25	4.5	22.5
work on methods	5	3	9	4.5	13.5	4	12	2	6
keep a good climate	4	4	16	2	8	4	16	4	16
organizational effectiveness	4	2	4	3	6	5	10	2	4
TOTAL			79		67.5		78		58.5

Legenda:

W = weight of criterium

A = Assessment of option vs criterium

Score = W x A

Fig. 5.14 Example of a decision grid as applied to improving a laboratory scientific production

suitable for keeping a good work climate, but less useful for working on methods and improving organisation effectiveness. The final choice was the better definition of roles and tasks as distributed among laboratory staff.

5.4.4   Example of Application: Contaminated Water in a Research Laboratory

A laboratory of a research centre experienced an abnormal rate of failures in micro-biology experiments. The problem was approached in a traditional way, however, for didactic purposes. Table 5.4 illustrates how it should have been managed

Table 5.4 A problem about demineralised water

Problem	Repeated failure of experimental procedures
Problem identification	Identify the type of failed experiments, localise the laboratories involved
Analysis of the problem	Quantify the percentage of past failures Identify batches of products used (reagents, chemicals ...) Control chemical contaminants and verify glass cleaning Control biological contaminants Check data with other laboratories and experiments Cross-checks between shelf life of chemicals, adequacy of lots, glass cleaning ... Make a hypothesis by a process of elimination: water quality Run experiments with various batches of distilled water from external suppliers Identify the cause: quality of the distilled water produced for the laboratory Verify the cause: water analysis => presence of contaminants in distilled water
Generation of solutions	1. Buy water from specialist supplier 2. Avoid possible contamination of distilled with demineralised water 3. Redesign hydraulic system of distilled water apparatus
Decision-making	Chosen solution: revision of the distilled water production
Plan and implement	Error-proof revision of the apparatus to avoid accidental mixing of demineralised and distilled water
Monitor and evaluate	Monitor the failure rate: no similar failure in the following 6 months

The table shows the reconstructed process of problem solving for the management of an abnormal failure rate in a series of microbiology experiments

according to the problem-solving wheel: which information should have been collected and analysed to find the cause, which options could have been considered, which ones chosen and implemented and finally which checks could have been identified and made to ensure a complete resolution to the problem.

5.4.5 Example of Application: Unresponsive Cell Culture

While testing biologically active molecules, a team encountered a problem with the cell lines expressing different receptors that were used to evaluate candidate compounds. The cell line was progressively reducing its responsiveness to the electrical stimulus, affecting the repeatability of the measurement, and thus the precision and accuracy of the output data. The problem was tackled with the aid of a fishbone diagram, filled in by the team during a brainstorming session. Potential causes were grouped into the classical 5-M categories of Ishikawa: man, machine, method, material and milieu (environment). Under the category *Man* the only potential causes identified were a lack of training and human errors. The category *Machine* regarded the components of the platform most likely to fall out of control because of their intrinsic complexity, namely stability of the laser, precision in the dispensing of the compounds, camera performance and control, and accuracy of the autofocus system. The *Method* was investigated by considering three main areas: the intensity and reproducibility of the electrical stimulus, timing of the various controls and procedure of cell washing before test. Regarding the *Milieu*, temperature, vibrations and stray light (i.e. from a source other than laser) were considered as potential causes. Finally, *Materials* included cells and (their) medium. Thus, the identified potential causes were investigated in more detail during the brainstorming and later by applying the 5-Whys technique. As an example, here is how one of the possible causes was explored:

- Why the response is not repeatable? Because the cells show different behaviour in different tests.
- Why the cells behave differently? Because they are subjected to mechanical stress during the treatment.
- Why they are mechanically stressed? Because they undergo transport, from the cell factory to the test laboratory.
- Why the cells need to be transported? Because cells are supplied by an external facility.
- Why the cells are supplied by an external facility? Because the test laboratory is not yet equipped for cell culture.

Eventually, dedicated tests proved transport to be one of the significant causes of cell stress and thus of the lack of repeatability of the measurement: the laboratory then decided to produce the cells directly on-site. All other possible causes were tackled to be proved valid or inconsequential.

Figure 5.15 shows an extract form the original and more complex diagram.

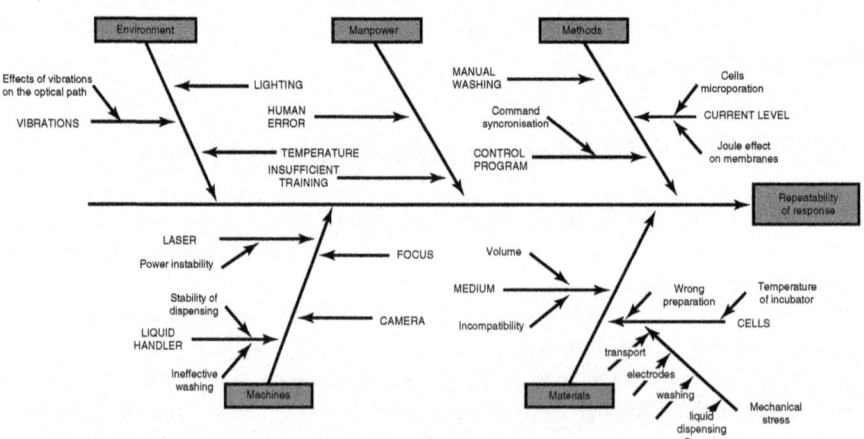

Fig. 5.15 Example of application of the fishbone diagram. The effect *Repeatability of response* was written on the head of the fishbone. The 5 *Ms*, methods, manpower, machines, materials and milieu (environment), were chosen to organise the causes by class/group

5.5   Knowledge Management

The process of creating, sharing, using and managing the knowledge and information of an organisation is referred to as knowledge management (KM). The KM allows to create added value to the organisation and enhance the current data culture. This allows a quick search for solutions or detailed suggestions, a faster resolution of operational problems, a tighter connection with colleagues and the generation of new knowledge. The KM is gaining recognition as an important element of business excellence or strategic programmes. The Malcolm Baldrige Award and the European Foundation for Quality Management (EFQM), the most important quality awards in the United States and Europe, respectively, recently added some criteria related to KM to their models for organisational excellence. The new ISO 9001:2015 among its innovations includes the clause 7.1.6, on organisational knowledge and its management.

KM needs a substrate both cultural and technological in order to be implemented. Top management and personnel should be aware of the importance of recognising, collecting and storing, for future access and sharing, all information vital to the core business of the organisation. A system, better if technologically based, should be in place for supporting storage and access to corporate knowledge. The importance of such management for a research institution, especially often subject to a high staff turnover, is unquestionable.

Before dealing with a few examples of knowledge management, we must clarify how data, information and knowledge are correlated, with the help of the *hierarchy of knowledge*, a model known as DIKW—which stands for data, information, knowledge and wisdom (Fig. 5.16). The first author to outline the knowledge management roughly in these terms was not a scholar but a poet: T. S. Eliot, in his play *The Rock*, wrote:

> *Where is the Life we have lost in living?*
> *Where is the wisdom we have lost in knowledge?*
> *Where is the knowledge we have lost in information?*

The model was built and expanded including *data* in further elaborations by several knowledge management scholars.

Data are discrete, objective facts relating to a phenomenon, i.e. elementary information units. If data are isolated, they have no meaning in themselves. They may be a number, an occurrence, an image or a sound. Examples are components and values measured in an assay, such as type of cell line used; concentrations of the different substances; number and timing of events; specific readouts, etc.

Information attributes meaning to data. The answer to the questions *who?, what?, when?* and *where?*, as applied to data, highlights the presence of relationships and thus construct information: who performed the test, when and where it was performed and, most importantly, how it was performed and what results were produced.

Knowledge: the question *how?* allows tracing or recognising a pattern within the body of information, by identifying events and cause–effect relationships.

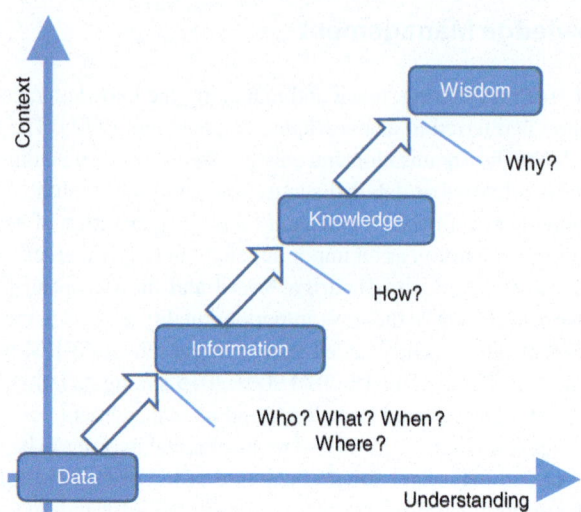

Fig. 5.16 The DIKW model for knowledge management. *X*-axis represents the *understanding* and the *Y*-axis the *context* or the ability to connect. Starting from the *data,* the four questions *Who?, What?, When? and Where?* allow organising them into *information,* increasing both the understanding and the ability to connect concepts. Asking *How?* highlights patterns and drives to *knowledge.* The full understanding of the phenomenon is reached by asking the last question *Why?* and this leads to *wisdom*

Accordingly, knowledge is essential to formulate a hypothesis about the functioning of biological processes, thus enabling intervention and their control.

Wisdom is the ability to develop, combine and use different knowledge to drive progress. By asking *why?,* the field of study can be widened to include the analysis of various and different phenomena, as well as the mutual exchange of information and experience from other fields: a specific model can be generalised, thus allowing a better understanding of the mechanism under investigation.

In summary, to become information, data must undergo elaboration and interpretation. Knowledge is reached when patterns in the information are recognised and understood. Wisdom is the deep understanding of the phenomenon, and the capacity to share knowledge.

The DIKW model is a good representation of how mastering the growing complexity (from data to knowledge) leads to an improved capacity of judgement, namely wisdom.

5.5.1   Technical Memory (Lesson Learnt)

Technical memory, namely the collection of all important information related to the (research) processes, can be seen as one of the elements of the organisational (or corporate) memory, the others being organisational knowledge, competence of staff, normative documents and any other data, information and knowledge collected within an institution.

Actually, technical memory is a set of structured information learnt from past scientific or technical experience, which can be used to consolidate and increase team's knowledge. It is also referred to as *lesson learnt*, to emphasise that it is a significant legacy from the scientific or technical past. It is of the utmost importance to store all competences earned from a project in an actionable way, in order to preserve and make them available for future use. This is crucial to research laboratories, where turnover of people is usually high.

More specifically, technical memory can be organised in a reference document repository used to store developed technical expertise as a set of files that can be browsed as well as searched. It can be implemented as a Web application and made available to all participants of a project or members of an institution, to be consulted when starting a new project or when dealing with problems to be solved.

5.5.2 Reports

So far, the importance of documenting the research activity has been clear and indisputable, not only for the final report requested by funding agencies or for publishing in a peer-reviewed journal, but especially for internal reference. Advantages are not limited to the *scientific memory* and the opportunity to collect data to facilitate the reproducibility of the study. The documented information (as ISO9001 calls the documentation) may be used also for student training and to disseminate scientific culture. Moreover, very often writing helps with understanding, clarifies thoughts and even fosters new ideas. For these reasons, the final report must be drafted immediately after the end of the study, and it should collect every aspect that is worth recording for future use.

GLPs require a final report on the study, signed and dated by the principal investigator and the study director, containing clearly identified headings:

- Identification of the study, test item and reference item
- Information concerning the sponsor and the test facility
- Experimental start and completion dates
- Quality assurance programme
- Description of materials and test methods
- Results
- Location(s) where the study plan, samples of tests, reference items, specimens, raw data and final report are to be stored

For a non-GLP research project, some of the topics above may be surplus to requirements, but the report should include at least (a) a clear identification of the project and the participants, as well as of materials and equipment; (b) methods and protocols for experimental activities and for data analysis and elaboration; (c) intermediate and final results, accompanied by information about raw data and archiving of samples; and (d) conclusions and forecasted opportunities.

The final report should be written by the principal investigator and approved by the scientific director, if appropriate or required. It should be shared with the research team, their supervisors and the sponsor if any.

5.5.3   Knowledge Sharing

Knowledge sharing is the activity through which information, skills and expertise are exchanged among people usually working in the same field. While knowledge sharing inside an organisation is operated by means of a KM system, it is not usual to observe it in the wide commercial world. The scientific world is different, because knowledge sharing is one of its basic principles and is achieved by the publication of scientific papers and scientific collaborations in studies and projects. Sharing knowledge through the publishing of articles is essential, but the information may not be organised. For example, there may be dozens of articles that deal with different aspects of the same topic, or with a subject that is progressively deepened in a sequence of articles; therefore retrieving complete information is difficult, decidedly time consuming or even impossible. In addition, non-scientific aspects of research management (e.g. precautions needed to manage a cell culture room, how to effectively conduct a certain type of assay …) are not considered worthy of discussion, and therefore are not easy to find nor formally credited. Whereas great attention has been dedicated to the sharing of knowledge in the form of guidelines in clinical or preclinical studies, non-regulated scientific research has not yet developed a common awareness of the importance of this topic. However, a structured system allowing researchers to share formalised information that is attributable and easy to retrieve would meet researchers' needs to access knowledge readily and effectively. Moreover, an enhanced uniformity among different laboratories would help to improve the reproducibility of the data and the exchange of material which would be treated and saved according to common instructions.

5.5.4   Example of Application: Quality-Based Scientific Guidelines—Definition, Collection and Sharing (the qPMO-WP1 Team)[8]

Guidelines are fundamental tools that provide valid indications for the proper running of a research laboratory, for the correct use of equipment and procedures and for aligning and standardising the procedures used in different scientific contexts. The identification, dissemination and application of common guidelines can improve significantly the reproducibility of the scientific results and the exchange of materials and data in the context of scientific consortia, also including industry partners and meta-analysis projects. One of the main tasks of the qPMO Network[9] has

---

[8] Giovanna L. Liguori, Giuseppina Lacerra and F. Anna Digilio.

[9] See Chap. 1/Acknowledgments.

been to focus on defining precise parameters for the drafting of guidelines, to avoid unnecessary steps and to maximise time and personnel resources. The outcome was a quality-based model for Life Sciences research guidelines, i.e. a canon for the drafting of guidelines, firmly based on both quality principles and methodologies, and documentation management [7]. The model has been developed by the qPMO-WP1 team at IGB and IBBR Institutes of CNR taking advantage of the plan-do-check-act (PDCA) cycle in order to identify all the activities necessary for the guideline definition process and inspired by the Ishikawa diagram (used for the first time in a guideline), to better organise the core theme of the guidelines (Fig. 5.17). At the core of the model is a flow chart describing all the steps to be followed writing a guideline: from selecting the drafting group and choosing themes, through to different levels of verification, and finally the publication.

The model has been followed by four different drafting groups located in different Institutes and Departments of the CNR, which in 18 months developed 13 guidelines. These guidelines are operationally focused, and cover main aspects of Life Sciences research from basic guidelines for laboratories to instrument management, facility management and research activities, as well as the application of quality methodology to research. Seven of these guidelines were used to provide the specific operating procedures for a QMS for a research laboratory (MarLab, IBIM CNR), certified according to the UNI EN ISO 9001:2008 standard in 2014 [8], giving further added value to the model. All guidelines have been applied by at least one Institute of the National Research Council. The definition of appropriate effectiveness and efficiency indicators for each guideline followed by monitoring activity (based on self-inspection, internal audit and feedback from researchers) showed a general improvement in research efficiency, such as in Drosophila, sea urchin and mouse housing as well as in cell culture facilities. The qPMO experience shows that the quality-based model for Life Sciences research guidelines identifies all the steps to be followed in the drafting and validation of guidelines and is applicable to

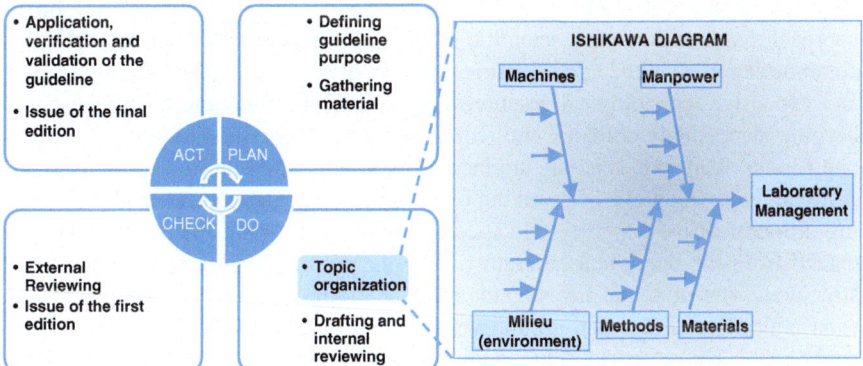

Fig. 5.17 Scheme for drafting a guideline. The procedure for drafting a guideline is organised following the PDCA cycle. The topic organisation in the DO phase is structured with the aid of the fishbone (Ishikawa) diagram with the 5 *Ms* methods, machines, manpower, materials and milieu (environment)

different scientific contexts and disciplines. Moreover, the flow chart describes simply and effectively the different steps, making drafting guidelines easy and immediate, as demonstrated by the high number of guidelines written by the qPMO group in 18 months of activity. This model can also be applied to define guidelines in non-regulated research laboratories that, on a voluntary basis, want to apply GRP as well as for technology transfer-oriented research and for the laboratories that aim to develop a QMS such as the ISO 9001.

The *quality4lab* Web Platform. All the guidelines provide a realistic scenario inside the institutes of the CNR and are published, maintained and updated in the *quality4lab* web platform[10] that has been created ad hoc by the qPMO-WP1 team, for collecting, cataloguing and disseminating best practices and scientific information among research communities. The CNR quality4lab website aims to be the expression of a virtual laboratory for life sciences where quality and scientific practices can come together to promote a culture of quality, knowledge sharing, procedural homogeneity and collaboration among scientists. The website strives to facilitate the cataloguing, sharing and consultation of scientific resources, focusing on protocols, model systems and molecular tools, available in CNR laboratories. Special sections are dedicated to the definition and sharing of best practice and to the models of application of quality management tools to research laboratory, developed and experienced inside CNR. Thus, this site offers researchers the opportunity to showcase their scientific activity, increasing its visibility and creating new opportunities for collaboration and networking, improving the knowledge and application of quality management to scientific laboratories and contributing to filling the gap between non-regulated basic research and applied research in which quality management is mandatory.

5.6   Team Work

Speaking about team work in scientific research is not trivial. The multidisciplinary approach required today by scientific studies entails networks of collaboration to combine expertise in the various fields of research and to provide access to complex and expensive technological resources. Moreover, the scientific disciplines have become increasingly complex and require skills too diversified to be found in just one person; therefore, scholars are increasingly cultivating narrow specialisations. A research project will be carried out by several working groups, each group bringing different skills, expertise and specialisations. The management will therefore extend from the individual laboratory activities to a network of partnerships and structures. The principal investigator must deal with a multisite, multidiscipline, even multicultural research group, which has to be effectively led to achieve the study goals. This scenario requires the principal scientist to be familiar with team dynamics and management, disciplines in which scientists have rarely received any training.

---

[10] quality4lab.cnr.it/en.

A team is a well-conceived, -led and -managed group of individuals that work *synergistically* to achieve a common target; synergy is the interaction of two or more agents or forces so that their combined effect is greater than the sum of their individual effects. Successful teamwork not only is the basis of the accomplishment of effective and efficient results, but also contributes to a pleasant, motivating and challenging work environment. Working in a good team lets people share competences and learn from each other, at the same time enhancing the abilities of the individual.

In managing the team roles and responsibilities, targets and tasks, resources, cooperation and consensus must be clearly identified and treated. Special care is to be devoted to the mood, maintaining a cooperative atmosphere and preventing, avoiding and properly handling conflicts. A team leader, as well as team members, needs the so-called soft skills, which require targeted training, and the provision of useful tools to foster teamwork. In the following section, some teamwork tools are illustrated, leaving dealing with team dynamics and psychology to dedicated literature.

5.6.1   Brainstorming

The technique of brainstorming consists of a brief session (20–30 min) of free contributions, aimed at collecting the largest possible number of ideas. Results of brainstorming can be processed by applying filters, such as constraints and preferences.

Brainstorming is widely used in problem-solving methods to find root causes (diagnosis), to identify adequate corrective actions (root cause analysis) and in risk assessment and management (planning and executing). It was devised in 1938 by Alex F. Osborn, an American advertising executive, who aimed at enhancing group creativity in dealing with problems, i.e. making each participant expressing their full potential. Osborne noticed that shy or discreet people are reluctant to share their contribution, because of the preponderance of more extroverted or assertive ones. He then decided to separate the two phases of creativity and result analysis, at the same time always favouring the production of large quantities of proposals.

Among variations developed during the last decades, the classical, simple brainstorming session needs a team leader to chair the teamwork and must follow a three-step format:

- Generation of ideas: The team leader presents the problem and the objective, together with any criteria or constraints to be considered. The team has a definite time (not more than 30 min) to express contributions. All people are asked for ideas, around the table or participating freely. In this phase, no one is allowed to make comments or criticise; ideas are combined and improved with the contribution of the whole team, and unusual ones are particularly welcome to stimulate creativity.
- Reduction of ideas: The team seeks a rational organisation of the ideas. It can be done by means of categories or filters; for example, *only the ideas that (a) maintain deadlines (b) must be under the direct control of the group and (c) should remain underneath the influence of the team leader.*

- Ranking of ideas (sometimes done by others than the team members): Every team member has a limited number of votes to choose the best ideas in their opinion. The most voted ideas are chosen to be further refined and are brought to a final decision.

In the second and third phases, some problem-solving/decision-making tools can be useful, as are the constraints window and the decision grid. The material from the brainstorming must be preserved as a source of further analysis, in case the ideas chosen and developed in the first phase did not prove satisfactory.

In recent years, some voices have arisen criticising the brainstorming technique. A psychology paper [9] states that in a brainstorming session human emotions and feelings, for example *evaluation apprehension*, could stress participants and harm the effectiveness of results. Moreover, according to the authors, exchanging ideas in a group reduces the number of domains of ideas that can be explored by participants: the participants' desire to belong to the executive core of the group may restrict, if not the breadth of ideas, the number of them that they may be willing to offer. The authors suggest taking breaks and summing up after a time, allowing the participants to think on their own to reach better solutions. Alternatively, the problem is presented before the session, inviting participants to find some solutions by themselves before coming to the brainstorming. In other words, getting a group of people to think individually about solutions, and then combining their ideas, could be more productive than getting them to think as a group.

On the other hand, a brainstorming session may help each team member feel that they have contributed to the solution, and it reminds each and everyone that others have creative ideas to offer. In this perspective, it can be a good exercise for team building, letting everyone experience cooperation and even some fun.

5.6.2   Holding Meetings

The meetings are a means of communication and management as well as productive teamwork, but it is wise not to overdo it. Well-planned and well-conducted meetings produce results and save time, while badly organised and managed meetings are a waste of resources, and hinder the participation of people in teamwork. The best measure of meeting effectiveness is if the meeting accomplishes its objectives to the mutual satisfaction of the participants, in the least amount of time. People come to meetings to get and share information, give ideas, make decisions and solve problems. In scientific research, meetings are especially useful for coordinating the different groups working on the same project and monitoring the progress of scientific studies.

Meetings should be planned and organised, but they may occur even standing in a hallway or gathering in an office. That type of meeting is sometimes more frequent, but if the members know how to communicate and act as a team, it can be as effective as planned ones. The key to meetings is high participation, effective communication and follow through.

The use of meetings as a way of communication should be covered by a *caveat*. The number of participants should be limited, involving only people who can contribute to the discussion of the topics on the agenda. Other interested people will be informed through the minutes. This consideration leads us to focus on two tasks: (1) planning and running of the meeting; (2) agenda and minutes: two paramount/ important tools for an effective meeting. A third one—debriefing—illustrated in the following dedicated paragraph can be used by the team leader to assess how the meeting was held and how the team is performing.

Meeting Planning and Agenda. Preparing the meeting is mainly a matter of defining the purpose of the meeting, the subject to be treated and the appropriate participants, together with date and place, the organisation of the meeting room (space, screen, projector and PC, flipchart, pens, everything that could be useful and refreshments if needed) and the drafting of an agenda. The team leader—who has called the meeting—should make a list of topics for discussion, with a planned duration for each one and possibly a ranking of priorities. The agenda should be sent to all participants a few days in advance, so that they may be informed on the subject and time allotted, agree on topics, effect possible changes in consultation with the team leader and prepare accordingly. An example of a simple agenda scheme is illustrated in Fig. 5.18.

Running of the Meeting. At the opening, the purpose of the meeting must be clearly stated and the agenda, and possible last minute amendments—especially the priority of topics—are illustrated and agreed upon. At the same time, it would be useful to choose among participants a *minute secretary*, in charge of recording considerations, decisions and actions taken during the meeting and drafting the minutes, as well as an appointed timekeeper—who could also be the chair or the team leader and helps in maintaining the discussion within the scheduled times.

The team leader, the principal investigator or whoever is holding the meeting should run the meeting according to the agenda and never let the group take the control. The *ground rules* are the perfect tool for this purpose. They provide a

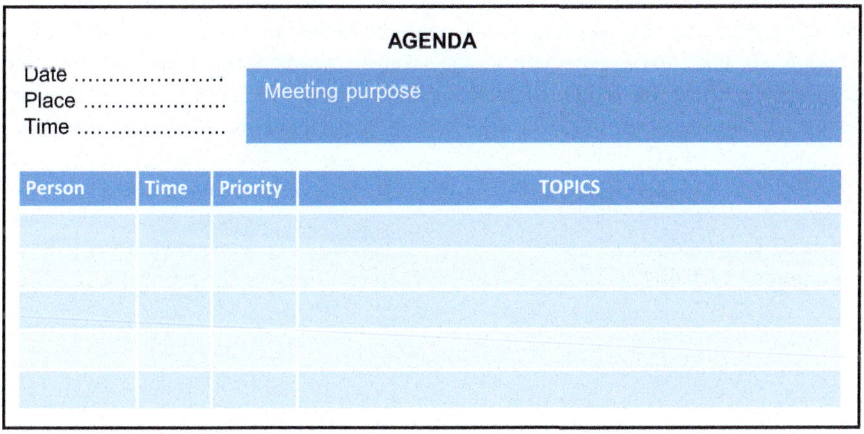

Fig. 5.18 Simple example of an agenda. *Date*, *place* and *time* of the meeting are recorded in the fields on the upper left. Meeting purpose is clearly defined in the upper field. The table lists who will make the intervention (*person*), the *time* slot dedicated to it, the *priority* agreed on the subject and the *topic* of the intervention

common style of how the group should function and help to regulate group behaviour. A simple set is set out below:

- Start and end on time.
- Everyone participates.
- Only one person speaks at a time.
- Stay on schedule.
- All ideas are group property.
- Separate the idea or comment from the person.
- Decide by consensus.

As can be easily seen, these rules are intended to set an open and neat discussion, prevent conflicts or interpersonal issues and nurture team spirit. Focusing on the respect of the matter and intervention time allows the meeting to be efficient and effective, saving time while achieving results.

Each team should agree on the list, adjust it to their own style of working and periodically review it. If the group behaviour is moving away from the expected, the team leader must remind all to respect the shared rules.

First topic of the meeting should be the review and follow-up of actions taken during the previous meeting(s). At the end of each item, the team leader should summarise what was discussed and decided, and agree on the actions to be taken with the responsible persons, within an agreed and defined time. If the discussion about one subject does not end within the allocated time, the group can decide to continue discussion—with an agreed modification to the agenda—or reschedule the topic for the next meeting. It is worth noting that respecting starting, discussion and closure times not only contributes to an effective and efficient meeting, but also gives self-confidence to the team, increasing motivation and commitment.

At the end of the meeting, key points for discussion and decisions should be highlighted and the list of actions reviewed to arrive at a shared conclusion. This being the case, the next meeting date is communicated or agreed. A short debriefing session can take place, especially if the group is newly formed, to give the team leader and participants an idea of the team performances. *Debriefing* can be achieved using a table (see Fig. 5.19) in which each participant in turn can express—in

Fig. 5.19 Scheme for debriefing. The first column is headed by a symbol representing the constructive aspects and is used to collect any positive comment by participants to the meeting. In the second column (denoted with a symbol representing the need for amelioration), suggestions for improving the running of the meeting are recorded

mandated order—one aspect as positive and the other as needing improvement (note, not *negative*). The debriefing results are quickly analysed by the team during a subsequent meeting. This arrangement gives the participants a chance to bring team dynamics into consciousness with enthusiasm and the determination to consolidate the group and increase synergy; the expression of emotions, doubts and difficulties is encouraged, so that they can be easily addressed and resolved to improve team cohesion and work.

Minutes of Meeting. The minutes of the meeting can be written by a recorder or by the team leader themselves, if they are skilled enough for both note-taking and controlling the meeting. The minutes of the meeting can also be drafted on a computer during the meeting and shown on the screen to have all the participants aware and in agreement on what has been annotated. A simple and clear *minutes of the meeting scheme* can be used, ideally divided into three parts: description of the meeting, summary of the topics covered and list of actions (past and present). Figure 5.20 shows a two-page form for meeting minutes.

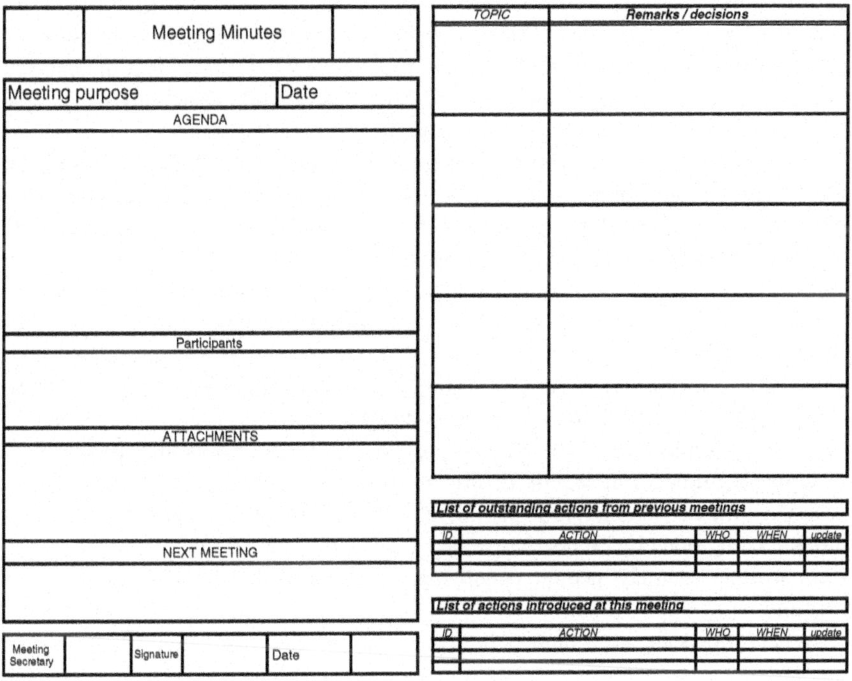

Fig. 5.20 Example of a minutes' form. The minutes of the meeting form is composed of two pages. *Date* and *purpose* of the meeting, *agenda*, *participants* and *attachments* are listed in the first one, together with the schedule of the *next meeting*. The form will be dated and signed by the *minutes secretary*. The second page is composed of three tables; in the first *topics* and main *remarks/decisions* are recorded. The two following tables collect the actions decided during the current meeting and those decided in previous meetings but not yet completed, also recording the person in charge, the expected date of completion and updates

Follow-Up. The team leader is in charge of ensuring that the minutes are compiled and distributed in a timely manner: 2 days after the meeting is a good turn-around time. They may also arrange for checks on actions, with brief calls or short meetings, to ensure that everything is ready for the next meeting, which, as always, should be well prepared.

5.7   Time Management

Time is a peculiar asset: it is unrepeatable, it passes and never stops and it is the decisive element, resources and results being equal.

Good time management allows to do the same things faster and to do the most important things first, thus increasing efficiency and effectiveness; in such way stress decreases, improving quality of life. In the following section, a few hints are given to deal with time management and the most common related issues.

The obstacles encountered when trying to best manage time can be both psychological and practical. Fatigue, concern and pressure are the psychological burdens that hinder the capacity of facing time management in the appropriate way: physical and intellectual fatigue are sometimes due to boring and repetitive activities; concern is generated by issues related to improvement and innovation, and anxiety about the outcome; pressure and stress are produced by the need to face unexpected events, urgent matters and possibility of chaos. Tired, worried and anxious people tend to procrastinate—and make mistakes.

To prevent psychological costs, strategies can be put in place: to prevent fatigue: do not put the same energy into every activity, have a short break every 60/90 min of work and spot and eliminate no-added-value tasks. Concerns can be limited by letting off anxiety: acting, working and living in time slots of 24 h, i.e. focusing only on deadlines and activities related to a day frame and planning the others, without chasing unrealistic goals, while managing risk. Pressure can be controlled by establishing deadlines for all tasks and setting priorities for action, and discharging nervous anxiety with physical activity, asking *what will actually happen if I do not do this* … and not worrying about unimportant tasks.

But which are the activities mostly responsible for time wasting? Rework, for example, the time spent to remake already (poorly) executed activities, and unnecessary work, i.e. making low- or no-value-added jobs or duplicating the work, already being done by others. A special category of unproductive time is linked to time wasters, namely people occupying our time with useless activities (e.g. chatting).

What follows are a few hints to help in the always challenging fight against time.

- Priorities should be managed in such a way as to complete most urgent and important tasks and not to struggle with other trivial matters. It is fundamental to strictly schedule activities, resources, deadlines and progress.

- People sometimes are found to exceed in accuracy well beyond the actual scope and need, for the sake of perfection: a *good enough* level—sometimes referred as *line of best fit*—must be established and achieved.
- Clarifying the preferred, most efficient and productive time of activity: in the morning or in the afternoon?—identifying this helps in concentrating at that time the most challenging activities.
- The last simple but important hint is to learn to say *no* to people trying to involve us in useless or time-consuming activities, as well as in jobs not matching our priorities.

There is a graduated approach in the management of time with the support of tools. The first step is the *To Do List*, a simple reminder of things to do. A better approach includes the definition of the objectives and the use of sticky notes and a simple calendar of events. Smarter time-managers plan activities on a daily/weekly basis with a view to manage and control priorities, having identified them and clarified values. Finally, the best way to improve efficiency is to use objectives and instruments that promote importance over urgency. The Eisenhower matrix is an example of such a tool.

5.7.1   Eisenhower Matrix

Dwight D. Eisenhower was a General in the United States Army and the Allied Forces Supreme Commander during World War II. He later became the 34th President of the United States, governing from 1953 to 1961. One of his most famous paradigms is *What is important is seldom urgent and what is urgent is seldom important.*

A correct approach to the Eisenhower matrix starts with the list of things to do. To be prepared to act with the adequate effectiveness, they have to be ranked in order of importance regardless of time, then the urgencies are identified and the scheme—Fig. 5.21—is completed. The scheme is a simple four-quadrant matrix:

- Upper right: DO NOW—urgent and important:
  - Face in the period of increased productivity
  - Divide the task into sub-elements
- Upper left: SCHEDULE—important but not urgent:
  - Plan, using diary/agenda and calendar
- Lower right: DELEGATE—urgent but not important:
  - Delegate to someone trusted
  - Ask collaboration from a colleague
- Lower left: DO LATER—not urgent nor important:
  - Put on standby, often vanishes
  - Trash

Fig. 5.21 Eisenhower matrix. Tasks must be put in one of the four sections, depending on their *importance* (related to outcome) and *urgency* (related to time). Tasks in the section *More Important/More Urgent* must be accomplished immediately. Tasks in the section *Less Important/More Urgent* may be delegated to a trusted person. Those in the section *More Important/Less Urgent* should be scheduled, while *Less Important/Less Urgent* tasks can wait and sometimes can be discarded

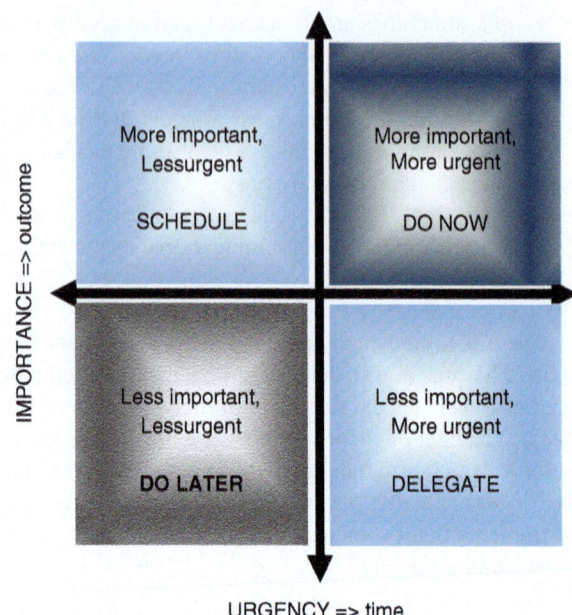

References

1. Hogarth RM. Judgement and choice: the psychology of decision. 2nd ed. New York: Wiley; 1987.
2. Tague NR. The quality toolbox. Milwaukee: ASQ Quality Press; 2004.
3. Lanati A. Qualità in Biotech e pharma: gestione manageriale dalla ricerca ai suoi prodotti. Milano: Springer; 2009.
4. Goldratt EM. It's not luck. Great Barrington: North River Press; 1994.
5. Annesley TM. Who, what, when, where, how, and why: the ingredients in the recipe for a successful methods section. Clin Chem. 2010;56(6):897–901.
6. Doggett AM. Root cause analysis: a framework for tool selection. QMJ. 2005;12:34–45.
7. Digilio FA, et al. Quality-based model for Life Sciences research guidelines. Accred Qual Assur. 2016;21(3):221–30.
8. Bongiovanni A, et al. Applying Quality and Project Management methodologies in biomedical research laboratories: a public research network's case study. Accred Qual Assur. 2015;20(3):203–13. https://doi.org/10.1007/s00769-015-1132-5.
9. Kohn N, Smith S. Collaborative fixation: effects of others' ideas on brainstorming. Appl Cogn Psychol. 2011;25(3):359–71.

Quality Methods for the Scientific Research

6

6.1 Project Management

Project management is an independent branch of knowledge, whose most important references are the Project Management Body of Knowledge (PMBOK®) [1], the ISO 21500 standard and PRINCE2. Those methods and principles are very useful also in the field of quality-oriented management systems, reinforcing the great attention quality plays to the correct and controlled development of any project.

Researchers from early in their career can be invested with the responsibility of defining targets and budget, leading, reporting and concluding scientific projects. Scientific studies are not the only challenging activity an investigator can be called to manage. A relocation of the laboratory, a consistent improvement action and the commencement of a new analytical method are few examples of activities that can be accomplished by means of project management guidelines.

In most cases, researchers are not trained in the field of project management. This specialized job, if not done methodically and based on appropriate knowledge, steals valuable time and resources to the experimentation activities, which may leave the researcher with an uncomfortable feeling of inadequacy. A method that provides templates, thought patterns and, above all, a clear understanding of concepts and distinctions helps in making tasks right from the start and in saving time and resources. These are the reasons for introducing a few concepts borrowed from the wider field of project management, without delving into it with the attention that this topic would deserve. In the following paragraphs, we will adhere to the method outlined by the PMBOK® of PMI and the ISO 21500 standard.

© Springer International Publishing AG, part of Springer Nature 2018
A. Lanati, *Quality Management in Scientific Research*,
https://doi.org/10.1007/978-3-319-76750-5_6

6.1.1   The Project and the Ten Knowledge Areas of PM

A *project* can be defined as a set of activities that achieves a specific objective, through a process of planning and executing tasks and the effective use of resources.

Unlike a process, a project has a well-defined beginning and end, and it is never replicated in the same way. The project manager (PM) is required to meet three main goals: time, cost and quality. *Quality* refers to the achievement of the target performances of the project, *time* refers to the timeliness of the tasks performed and *cost* is the compliance with the budget, which involves careful management of resources. Managing a project needs skills in many different areas, not only quality, costs and time. The PM has to deal with people, suppliers, communication and risks and keep it all together in an organic whole. The PMBOK® and the ISO 21500 standard [2] divide the project management into ten areas of knowledge:

1. Project integration management
2. Project stakeholder management
3. Project scope management
4. Project time management
5. Project resource management
6. Project cost management
7. Project risk management
8. Project quality management
9. Project procurement management
10. Project communications management

The PMBOK and the ISO standard identify also five processes to manage effectively each of the knowledge areas:

1. Initiating
2. Planning
3. Executing
4. Controlling
5. Closing

As an example: for the *project integration management* area of knowledge, the initiating process is *develop project charter*, the planning process is *develop project management plan* and the executing process is *direct and manage project work*; there are two monitoring and controlling processes, namely *monitor and control project work* and *perform integrated change control*, while the closing process is as expected *close project or phase*. The knowledge area processes differ slightly between PMBOK and the ISO standard. As one might expect, an in-depth discussion about project management would be lengthy and complex; therefore only some of the most useful concepts will be highlighted.

6.1.2 The Targets

Not only the entire project, but also every single task needs setting a target. Defining a target could appear an easy, straightforward job for a research project, but it can hide some pitfalls; the better it is defined, the easier it is managed. By using an acronym, a well-defined target must be SMART:

- Specific: clear and unambiguous statement
- Measurable: having quantifiable characteristics
- Achievable: realistic
- Relevant: important and adequate to the research
- Time specific: referring to a well-defined date or a specified amount of time

These characteristics emerge from simple considerations. A target must be clearly described to avoid misunderstanding among participants to the project, duplication of efforts on the same job and leaving grey areas unattended. A target cannot be considered as having been reached, and to what extent, if it has not been defined in a quantifiable manner. There are few situations more demotivating for the participants than being recruited for an impossible mission, towards an unreachable goal: targets must be achievable and compliant with available resources. Engagement and efforts will be more significant if the target of the project is important, for the laboratory itself, for the institution and, in general, for the science. Finally time constraints are important: there is a significant difference in setting targets of a few months or a year or more, as resources, commitment and feasibility will be affected by time constraints.

For a sound definition of project objectives, there is the need for a clear understanding of the present situation *as is* and for a well-described future situation *to be*. The following suggestions can be of help in streamlining and organising the project's objectives in a clear description: starting from a simple investigation that can involve interviews with researchers and management to more structured tools such as Ishikawa diagram, gap analysis, brainstorming, decision grid and PDCA, just to mention some examples.

6.1.3 The Project Charter

A project charter is a formal document used in project management to describe the purpose, context, resources and expected results of the project. It is the reference document for the organisation and the team, and it is used to build a common target and scope for all participants.

A project charter should specify at least:

- Project scope
- Targets and related assessment criteria
- Skills needed

- Resources
- Reporting and reviewing
- Terms of execution and completion
- Composition of the project team

The usefulness of describing a project with a charter is not limited to a project proposal for funding, but its worth becomes evident also when managing internal scientific or improvement-oriented activities. To draw up a charter, it is useful to refer to a simple scheme like the one known as the Kipling's servants, which covers all major subjects to be described.

6.1.4 The Work Breakdown Structure

The PMBOK® describes the work breakdown structure (WBS) as a deliverable-oriented hierarchical decomposition of the work to be executed by the team. The WBS is a method to deploy graphically and hierarchically the scope of the project in the tasks needed to complete it. Besides the immediacy of a graphic tool, WBS has several advantages which allow to improve the estimation of resources, to control the execution and progress of the project and to verify its completion more accurately. In most organisations, the WBS can be defined as a general template, ready to be customised and filled in greater detail for similar, future projects.

To build a WBS, the project must be divided into areas, and then each area is further divided into lower level tasks and objectives. The outcome is a tree-shaped chart whose lower components are the *elementary activities*, often called *working packages*. Each activity can be described with its target, deliverables and required resources. The deployment of activities streamlines the drafting of the project schedule and the evaluation of resources: a resources plan can then be drafted summing up the required resources for all the elementary activities.

6.1.5 The Schedule

A schedule is a listing of the project's milestones, activities and deliverables (see Box 6.1), with defined start and end dates, resources and personnel in charge. Graphical tools are always used to draft a project schedule; even for very simple ones, a table listing activities and milestones, personnel in charge, start and end dates and expected deliverables (if not listed among activities) can be adequate—in this case the need for resources should be described in another document.

A schedule is vital to plan the project and to check its progress. Moreover, among the planned activities, some control points should also be inserted in order to check both progress in the results and timeliness in the execution of the task, and to correct them if needed. Project reviews[1] are just one example of such control points.

---

[1] See Chap. 4, Sect. 4.5.2.2.

Among graphic tools, the most used are network diagrams, PERT and Gantt charts. For complex projects, project management suggests to first draw up a network diagram and then a Gantt; the reasons are illustrated in the following section.

Note: Specific software products to design and manage planning of activities, resources and constraints are commercially available.

Box 6.1: Activities, Deliverables and Milestones

Researchers are not always very familiar with these terms that are widely used in project management. Here are definitions and some simple examples:

Activity: An individual task that must be performed and completed in order to generate a result (*deliverable*). An experimental phase—e.g. characterisation of a specific protein—is an activity.

Deliverable: The result of an activity or an entire project, usually released to the final recipient (e.g. partner, customer). A deliverable may be a tangible or an intangible good. For example, data and information from an experimental phase or production of a purified protein are the deliverables of an activity within a project. The confirmation of a scientific hypothesis, the publication of a paper and the study report may be the deliverables of an entire project.

Milestone: A major accomplishment marking a specific point along the project timeline. It is often referred to as an *event*. In a granted project, periodic reports to the funding agency are milestones, as well as the dates at which intermediate products (e.g. the produced purified protein) are provided to a laboratory partner in the project.

Network Diagram/PERT. PERT—short for program evaluation and review technique—was developed in 1950s in the US Navy environment. It is a useful planning tool for projects where activities have complex dependencies, because it graphically describes the logical and temporal relationships. It is also known as network diagram, because it allows to examine a project as a series of events and actions linked to each other in a network of relationships.

To draw the PERT diagram, targets that interact with each other and the tasks to achieve them must be defined. Three forecasted time intervals (optimistic, probable and pessimistic) may be identified for the attainment of each target or subproject. The project development is illustrated by nodes (rectangles, diamonds or circles) representing the individual tasks or events (i.e. null duration tasks) and by lines representing the logical and temporal connections. For each task, the duration or the start and end dates are defined.

The graphic description of logical and temporal dependencies allows the identification of the so-called *critical path*: it is the sequence of activities that determines the minimum duration of the project and therefore the success or failure of planning: a delay on the critical path affects the overall project timing. Figure 6.1 shows a very simple network diagram and a more complex one for the implementation of

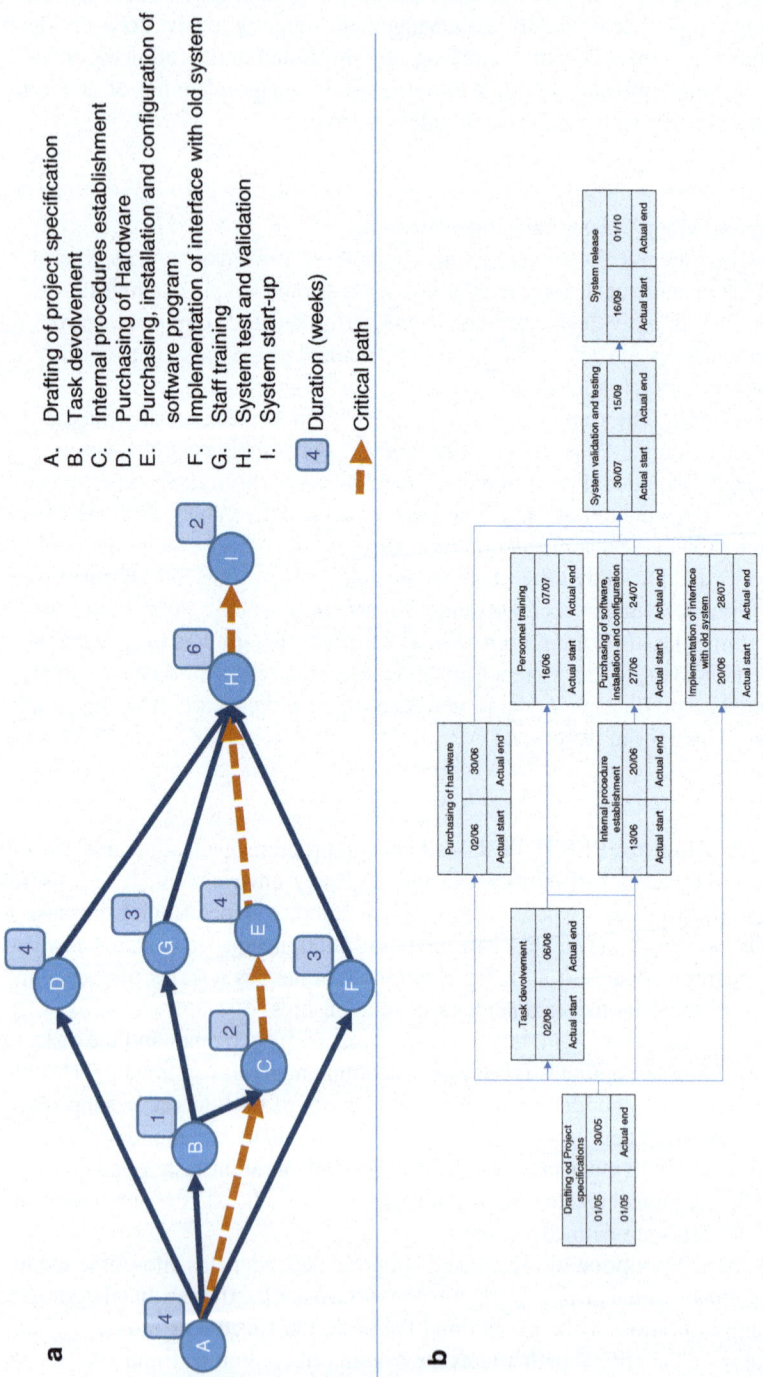

Fig. 6.1 Examples of network diagram/PERT chart. Both diagrams are about the implementation of the same hardware/software programme. (**a**) A simple network diagram with expected duration for each task and the critical path highlighted. (**b**) PERT diagram with expected dates. The fields for the actual dates are left blank for tracking

a hardware/software programme, where expected duration (a) or dates—leaving blank the fields for tracking actual dates—(b) are planned. The critical path is also highlighted.

PERT allows also to plan for the allocation of resources (equipment, man/hours, access to laboratories, etc.) and verify their use in defined time frames. The resources of a laboratory may be used on different projects, for limited periods and with different involvements. A PERT, that indicates for each task and each project the need for resources, helps in maintaining under control where, how much and for how much time each technician or researcher is required for the various activities and to avoid clusters of requests or unmanageable overlaps.

Gantt Chart. A Gantt chart (also known as bar chart) is a simple tool to plan a project: it is named for the American engineer Henry Gantt, who devised it in the early 1900s. Frederick W. Taylor introduced it in scientific management as early as 1917. On a chart, time is shown along the X-axis, and each activity is represented as a bar or a line. This simple representation shows any sequence of operations in time, allowing to check the progress and to verify the degree of completion of the project. The Gantt chart is very useful for rather simple projects, whose phases are not interrelated in a complex way by logical or time dependencies.

Figure 6.2 illustrates the Gantt chart for the same project as in Fig. 6.1.

As seen when discussing the planning of a research project, relationships among the various activities should be integrated in the plan, to avoid inconsistencies, conflicts and uncontrolled impacts on the whole schedule, possibly due to difficulties and delays with a task. The Gantt of Fig. 3.7 illustrates the planning of a study with some dependencies that can be better shown and managed with a slightly different representation. While the Gantt chart does not show logic and temporal dependencies, the network diagram and the PERT chart allow a more comprehensive outlining of times, but do not show the personnel in charge of each task. A better solution is illustrated in Fig. 6.3, where both personnel in charge and mutual dependencies can be clearly shown. Most software programmes that support automatic scheduling adopt this representation.

6.1.6   The Management of Costs

One of the most challenging tasks for the manager of a funded project is managing the budget, i.e. the amount of money devoted to the development of the research project. The budget is a detailed operational plan for the economic management of the project within a defined period of time. It is not a forecast, but a programme establishing the commitment of the person in charge to obtain effective and efficient performance (use of materials, cost containment, etc.). Often it needs the support (within or outside the team) of experienced staff, but it is always wise for the principal investigator to have some knowledge about the basic mechanisms of budget management.

First, the period of reference must be defined: it can be a fixed period of time (a year), the duration of the project or of each project phase. Then the project costs must

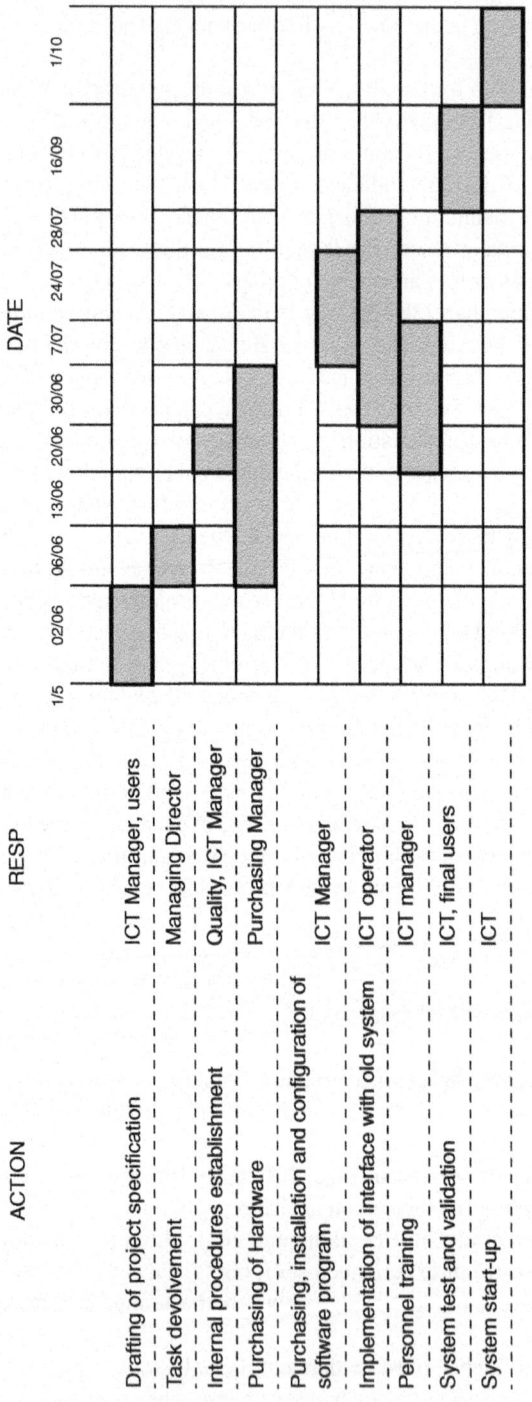

Fig. 6.2 Example of Gantt chart. Actions are listed in the first column, personnel in charge in the second one and the matrix (dates in column, actions in rows) shows when the actions must be executed

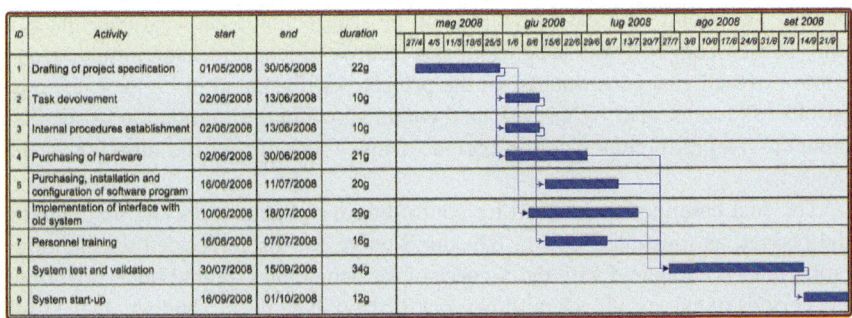

Fig. 6.3 Example of Gantt chart with mutual dependencies. For the same project of Fig. 3.21, this type of Gantt chart shows the mutual dependences between tasks, as lines that link preceding to succeeding tasks

be estimated before bringing the project to approval. The total cost of the project is determined by means of an estimate, while the budget defines when and for which reasons expenses are made. This is a means of keeping performance and economic management under control. Costs can be divided into four simple categories: staff (man hours, research grants, scholarships …), materials (cell lines, reagents, consumables), expenses (travel expenses, conference fees …) and investments (equipment, tools). Special care is to be devoted to the purchasing of equipment and other goods whose use can be spread over several years. In such a case, the total cost must be divided by the number of years of use and only the portion of cost related to 1 year of use must be attributed to each year. This mechanism is called *amortisation*.

Other classifications, useful to properly attribute costs, are into fixed costs and variable costs, as well as in direct and indirect costs. In the first case, variable costs depend on the volume of activities (e.g. work hours, machine hours, materials) while fixed costs are influenced by time but not by volumes (e.g. energy, rent, insurance). In the second case, direct costs are attributable to employment factors, assigned to specific deliverables, and indirect costs are those not directly attributable to the project (e.g. overheads).

Cost estimation may be made by a comparison with similar, past projects, or using standard parameters provided by the institution or developed in the laboratory. In any case, the best way is to take as a reference the WBS of the project and attribute the different kinds of direct cost to each WP. The evaluation can then be completed by adding any foreseeable indirect cost. The main project management standards suggest allocating some buffers to cover risk controls (*risk budget*) and change management (*change budget*), as well as a *contingency reserve* to cope with uncertainty (under the responsibility of the project manager) and a *management reserve* at the disposal of upper management.

For each estimation, data source, method/criterion used and its reliability must be known and should be recorded. Distribution of the costs throughout the duration of the project is then planned, being informed by the WBS, the project schedule and the latest estimation of costs subdivided into working packages and time. Project management uses the *cost baseline*, which is a useful diagram for depicting the

budget, and monitoring cash flow (as will be seen below). The cost baseline is a
time-phased budget that is used as a basis against which to measure, monitor and
control overall cost performance on the project. As illustrated in Fig. 6.4, the cost
baseline is represented by an S-shaped curve, showing the summation of the discrete costs—per months or other period of time—throughout the duration of the
project.

The cost baseline can be used for controlling the expenditures and maintaining
the budget, as illustrated in Fig. 6.5: the S-curve for the forecasted expenditures
(solid line) is compared with the S-curve of the actual ones (dotted line), highlighting periods of excess of cash outflows and drifts with respect to budget. At the time
T_1 the expenses on the project are less than foreseen, but at the time T_2 the expenses
are exceeding the budget and actions are required to limit the extra-budget costs. On
the same diagram, the green line indicates the available amount of cash—for example for funding payments—for the project throughout its development. The comparison between the green line and the S-curve highlights possible problems such as
when an expense is required but there is no cash available, a situation to be avoided
when planning the expenditure.

Fig. 6.4 S-curve (cost
baseline). The S-curve is a
plot of costs against time.
The *dotted line* shows the
single costs (i.e. the
discrete costs), while the
solid line represents the
sum of costs at a specific
time (cumulative costs)

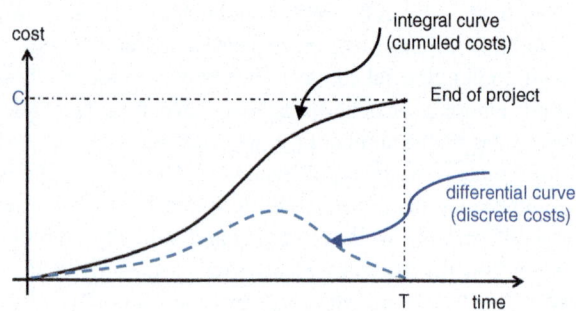

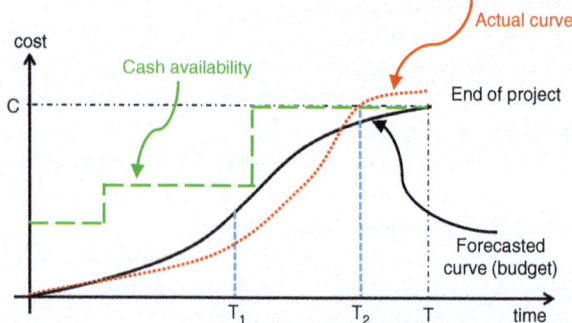

Fig. 6.5 Controlling the budget. The S-curve of the forecasted costs (*solid line*) is compared with
the S-curve of the actual expenditures (actual curve, *dotted line*) and with the cash availability
(*dashed line*). At time T_1 actual costs are lower than forecasted costs and there is good cash availability, while at time T_2 actual costs are higher than both expected costs and cash availability, causing a financial problem for the project

Budget management as well as control-criteria must be accurately planned during the project scheduling and periodically carried out. For example, one lab meeting out of four may be dedicated to the monitoring of costs (and risk). Variations to the budget may be due to unforeseen issues such as technical problems, difficulty in accessing or operating the instrumentation, changes to the objectives or lack of supplies. To these causes, some others can be added, such as an incorrect evaluation of the maturity level of the project or of its complexity, lack of risk assessment or underestimation of cultural and environmental factors. Whatever the nature of the problem is, corrective actions are required. They can range from a modification of the project schedule and consequently of the cost baseline, to a serious recovery plan.

6.1.7 A Simple Example of Project Planning

Figure 6.6 illustrates a simple example of a WBS for a scientific project consisting of two distinct phases of evaluation, in vitro and in vivo, for the activity of a substance.

For each *working package* of the WBS (a) human resources, material, expenses and equipment are evaluated. The network diagram (b) allows to highlight the mutual dependencies—in this case tasks 1.2 and 1.3 can run independently, task 1.4 cannot start before the end of tasks 1.2 and 1.3 and tasks 2.1, 2.2 and 2.3 are linked to the predecessor. The WBS and the network diagram then make it easy to draw the timeline of the project (Gantt chart, (c)). Resources can then be added to include the specific amounts required in each phase throughout the project, and split in the actual needs along the timeline. At the bottom of the diagram the *resource plan* (d) for human resources (units 1–3) is derived from the WBS, the network diagram and the Gantt chart. Similar resource plans can be drawn for material purchase, expenses and use of equipment.

Figure 6.7 shows the cost baseline for the sample project described in Fig. 6.6. Forecasted costs are attributed to each work package and slot of time (month), and then calculated for each month. The points of the cumulative (S-shaped) curve are computed adding the sum of costs of previous months to the amount of each month, and the curve is then drafted (solid line). The red (dashed) line refers to the actual expenses and the green (dotted) line to cash availability. As it can be easily seen, the project has several critical points: starting from month 5 cash flow is negative, i.e. expenses are higher than cash availability; at the same time actual costs are overtaking the forecasted ones.

6.2 Failure Mode and Effect Analysis

FMEA is a risk analysis method applicable to a process or a project at various levels of detail or feature (system, product, service and function). Born in the 1970s in the aerospace field (NASA) for the control of project and process risks, it spread

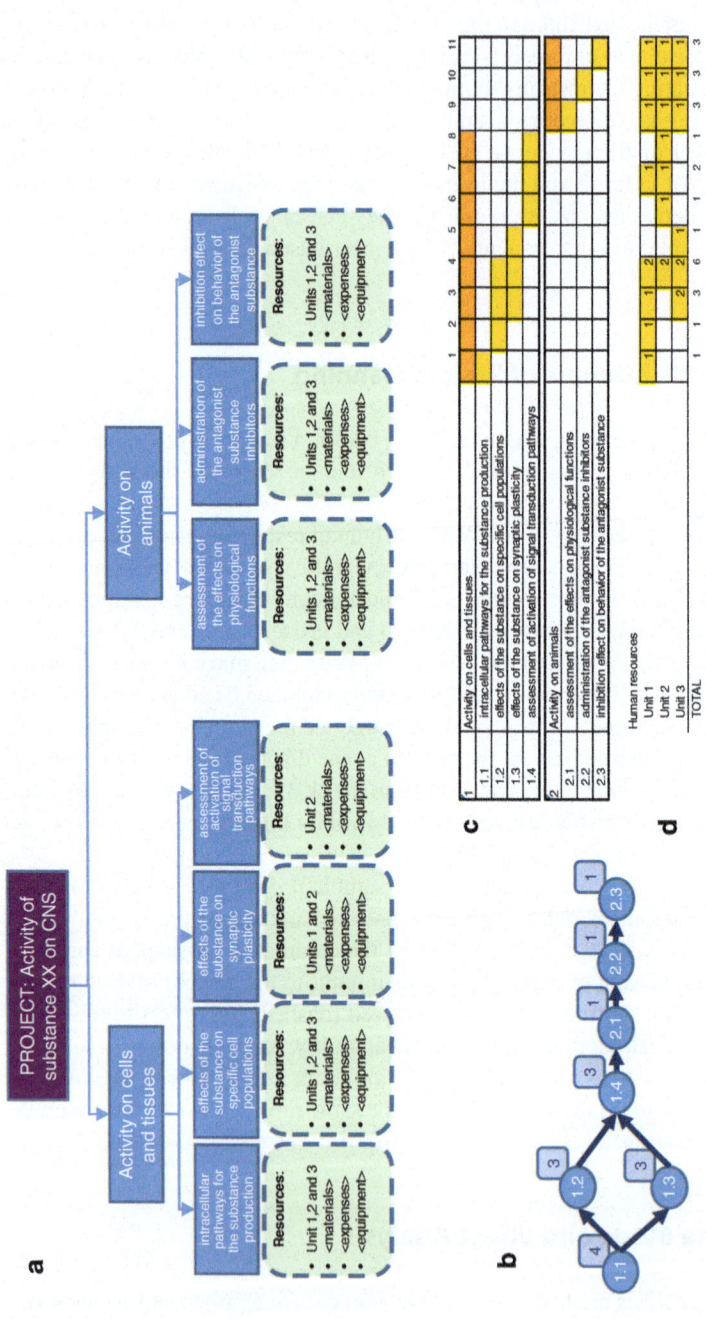

Fig. 6.6 Example of WBS, network diagram, Gantt chart and plan for resources. (**a**) The WBS for a scientific project regarding the activity of a substance consists of two distinct phases, for in vitro and in vivo evaluations. The need for human resources, material, expenses and equipment is estimated for each *working package* of the WBS (*dashed boxes*). (**b**) The network diagram shows mutual dependencies between activities as well as their duration. (**c**) The timeline of the project (*Gantt chart*) is drawn combining the information of WBS and network diagram. (**d**) The *human resource plan* is built on estimating which resources are required in each time slot throughout the project, by getting information from the WBS, network diagram and Gantt chart

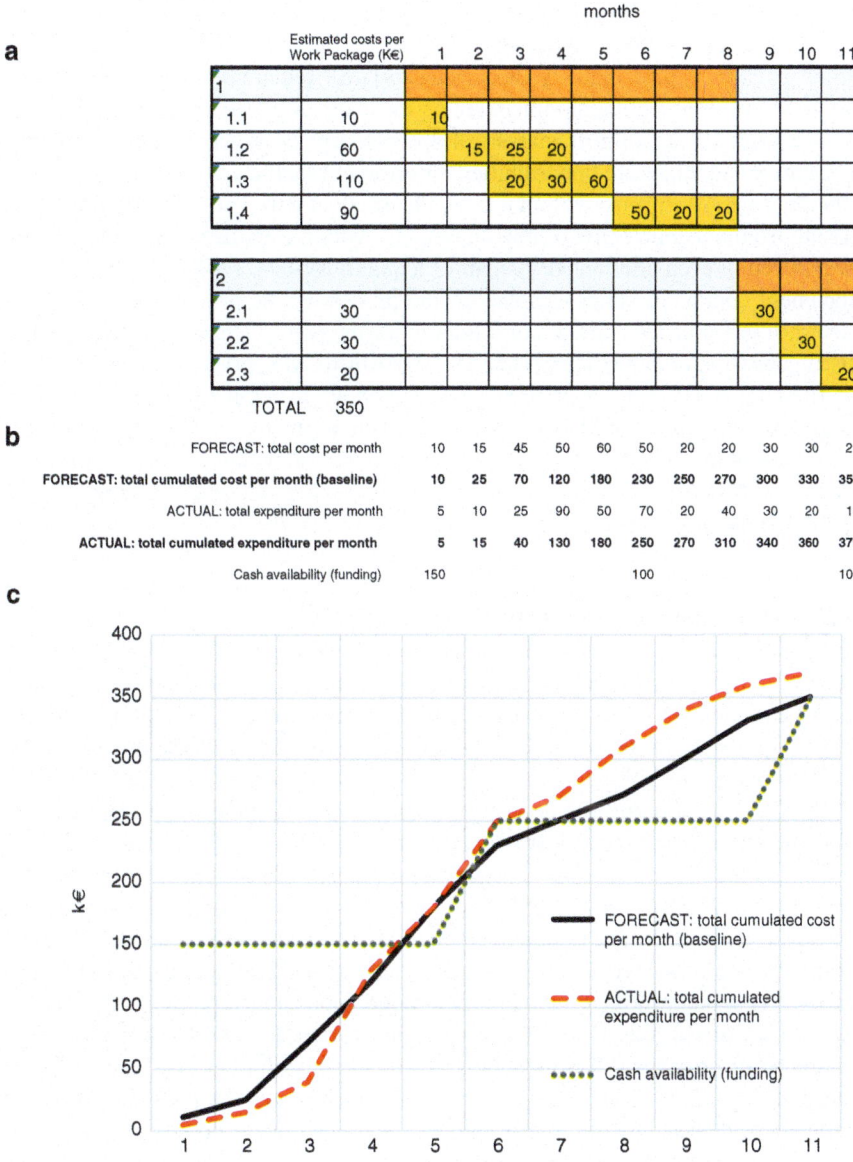

Fig. 6.7 Example of budget control with the cost baseline (S-curve). For the project of Fig. 3.25, lasting 11 months, the following are shown: (**a**) the resource plan with the estimated costs per work package. (**b**) The calculation of cumulative *forecasted costs* and the record of discrete and cumulative *actual expenditures*, as well as *cash availability* from funding. **c**) The cumulative values plotted on a graph: cash availability (*dotted line*), S-curve of budget (*solid line*) and actual expenditures (*dashed line*)

rapidly, especially in the automotive industry. The last years have seen a widespread use of FMEA for different products (software and services) and new fields, mostly in the pharmaceutical industry (according to GMP requirements) and healthcare, where it has been slightly modified, taking the name of HFMEA.

The FMEA is an a priori analysis of process steps or project function, intended to identify and eliminate opportunities for error and their causes, in order to maintain the process or project within acceptable limits of risk. The costs of corrections arising from nonconformity or non-compliance with the end-user requirements can be reduced or even eliminated. As part of a quality system, FMEA can be a tool to document continuous improvements to the process or the product, becoming the record of experiences and the consolidation of the *lesson learnt*. The most recognised general international reference for FMEA is the IEC 60812 standard.

To maintain a simpler approach to the methodology, in this text the discussion is limited to the process FMEA or P-FMEA, leaving to the existing literature the details about the design FMEA or D-FMEA. In the following section, some examples of P-FMEA are illustrated as applied to protocols in molecular biology or biotechnological production.

6.2.1   Methodology and Tool

In order to provide competent and complete contributions to the analysis, the working group involved in the development of the FMEA should be multidisciplinary and composed of all personnel involved in the development of the process under analysis.

First, the process must be described by means of simple steps or sub-processes, using flow charts to have a clear overview of individual operations and to facilitate understanding. Then the team identifies all possible *failure modes* within the process, potential *failure causes* for each of them and all the possible *failure effects* on the final user or result. The evaluation of these three aspects is based on data obtained from previous experiences, from literature and from the skills and understanding of the experts in the working group. For each failure mode, the importance of the effect on the user or the final result is assessed by a parameter called *severity* (S). The parameter **occurrence** (O) is then used to rank the probability of occurrence of the cause of the failure mode and the *detection* (D) estimates the possibility that the error can be detected by the controls already envisaged by the process. Each parameter is given a numerical value, usually between 1 and 10, referring to specific tables: the value scale is defined depending on the accuracy required for evaluation and the experiences or data available. These assessment tables are the result of the field of application: in more established ones, they are detailed and complex, resulting from the analysis of different aspects of the process and from the experiences in process and product. Tables 6.1, 6.2 and 6.3 illustrate three simplified ranking scales that can be used as a first reference.

The construction of the assessment tables is a key point for the analysis process, since it is the only way of detecting and encoding the fault conditions. It is important to provide a brief and clear description for each point of the scale, to share the

Table 6.1 FMEA severity

Severity rank	Description
10	Hazardous, without warning
9	Hazardous, with warning
8	Very high
7	High
6	Moderate
5	Low
4	Very low
3	Minor
2	Very minor
1	None

Simple ten-step scale for assessing the severity of the effect of a failure/error on the final result of the process under analysis

Table 6.2 FMEA occurrence

Occurrence rank	Description
10	>100 per 1000
9	50 per 1000
8	20 per 1000
7	10 per 1000
6	5 per 1000
5	2 per 1000
4	1 per 1000
3	0.5 per 1000
2	0.1 per 1000
1	<0.01 per 1000

Simple ten-step scale for assessing the probability of a specific cause related to a failure/error in the process under analysis

Table 6.3 FMEA detection

Detection rank	Description
10	Impossible
9	Very remote
8	Remote
7	Very low
6	Low
5	Moderate
4	Moderately high
3	High
2	Almost certain
1	Certain

Simple ten-step scale for assessing the coverage of controls aimed at identifying failures and/or negative effects of failures in the process under analysis

same understanding among the working group, and to give proper weightings with a view to determine the need for corrective actions.

It is worth noting that the scales of the tables for severity/occurrence and detection have an opposite trend, from more (10) to less (1) severe in the first case, and from least detectable (10) to the most easily (1) detectable in the second case.

The risk priority number (RPN) is then defined as the product of the three parameters S, O and D. It allows making decisions where improvement actions are needed. A numerical limit for RPN, meaning the maximum permissible risk, is seldom defined to allow identification of those failure modes deserving preventive or improvement actions: choosing a low RPN threshold means determining the most ambitious actions to reduce opportunities for error, making the process more robust and reliable. However, recent international references for FMEA [3–5] suggest not using a RPN threshold to determine the need for actions, because it can hide specific cases characterised by very high severity, but low occurrence and detection. Moreover, as pointed out by the AIAG[2] guidelines [4], *establishing such threshold may promote the wrong behaviour causing team members to spend time trying to justify a lower occurrence or detection ranking value to reduce RPN* avoiding to address the real problems. The guidelines suggest simply addressing errors with the highest value in one of the three parameters S, O and D, possibly applying the *80-20 principle* (or Pareto principle). Taking this course—i.e. Pareto principle—requires defining the actions necessary to lower the first 20% of failure modes ranked by the higher RPN, not forgetting that special attention should be given to failure modes with a high severity.

The actions identified may have an effect on the human, technical or organisational factors, and must be properly planned. The three parameters S, O and D, based on the results after the identified interventions, are then re-evaluated and a new value for RPN is calculated. Should the RPN value remain unacceptably high, new interventions should be considered to achieve an acceptable result. The FMEA must always be updated if either the process undergoes changes or new information becomes available.

For the application of FMEA, a worksheet such as the one shown in Fig. 6.8 can be employed. The first columns are used to identify each failure mode, determine the effect and related S, record the causes and their O and identify controls already in place assessed by D, as well as the resulting RPN. The next columns are dedicated to actions identified by the team for keeping the process under control, the person in charge of the actions and the deadline. Finally, the last four columns include S, O and D parameters with the new RPN recalculated in the light of the actions introduced.

---

[2] Automotive Industry Action Group: AIAG.org.

→

Fig. 6.8 Scheme for the FMEA. In the chart for the FMEA analysis of a process, the first information to be recorded is the process name and owner (upper left), the team (field Prepared by, upper right) as well as date of completion and revision. The first two columns are dedicated to the description of the process step and possible failures/errors. In the next six columns are recorded the effects on the final user, the causes of failures and the controls together with their respective assessments in terms of severity, occurrence and detection. The column RPN collects the product of S, O and D for each failure identified. The following three columns are used to plan preventive actions identifying the person in charge and the expected date of completion, for every process step needing improvement. The last four columns are devoted to the recalculation of S, O and D and the final RPN

Failure Modes and Effects Analysis

Process or Product Name:			Prepared by	Pages:
Process Owner:			FMEA Date:	Rev.

Key Process Step or Input	Potential Failure Mode (Error)	Potential Error Effects	S E V	Potential Causes	O C C	Current Controls	D E T	R P N	Actions Recommended	Resp.	Actions Taken / date	S E V	O C C	D E T	R P N
What is the Process Step?	In what ways can the Process Step fail?	What is the impact of the error on the Output (customer or internal requirements)?	How Severe is the effect on the Output?	What causes the process to go wrong?	How often does Cause or Error occur?	What are the existing controls and procedures that prevent either the Cause or the Error?	How well can you detect the Cause or the Error?	SxOxD	What are the actions for reducing the occurrence of the cause, or improving detection?	Who is Responsible for the recommended action?	Note the actions taken. Include dates of completion.				

From a quality point of view, FMEA is a continuous improvement plan that evolves with the process or product analysed. The team becomes accustomed to observe the processes, freeing the analysis from personal bias, without any concern and bringing it to the attention of the personnel in charge. Another advantage is the establishment of a reservoir of experience to be drawn upon for the management of the process, and especially for the design and industrialisation of new processes. In this manner, the transfer of experience and knowledge to a new process takes place in an analytical and a controlled way.

6.2.2   Example of Application: Analyse the Risk in a Biotech Production Process (Marta Galgano[3])

The AFI[4] working group on Biotech and ATMP Medicinal Products performed a FMEA analysis of an advanced therapies medicinal product (ATMP) manufacturing process, in order to test the technique in this specific field. The choice fell on a standard cellular production growing in adherence, already validated and suitable to be used for human use, i.e. approved for clinical trials and manufactured by an authorised GMP facility. Figure 6.9 illustrates the process flow chart.

Each step of the process was analysed by the working team, using a standard FMEA scheme (Fig. 6.8) and the general criteria described in the previous paragraphs. Slightly different S, O and D tables were used (Table 6.4). For the demonstration purpose of the analysis, the working group decided to apply the RPN threshold, which was set at the value of 100.

The seven process steps were analysed in depth, identifying the most probable problems, related causes and controls for each step. Figure 6.10 illustrates an excerpt from the complete FMEA analysis table: the operation considered is formulation and filling. Three possible problems were detected, two of them having two possible sources of error each. For the two potential causes *error in sampling* and *complexity in aliquot for product characteristics* the controls already in place did not show the required effectiveness, causing RPN values above the set threshold of 100. Thus, two actions (reported among the others in Table 6.5) were required to lower the values of occurrence or detection.

Figure 6.11 summarises the RPN calculated before and after the implementation of preventive actions.

---

[3] Project & Quality Manager at Blast Research, and Working Group Coordinator at Associazione Farmaceutici Industria.

[4] AFI (Associazione Farmaceutici Industria—Scientific Society) is a cultural, professional and scientific association of graduates in scientific disciplines, drawn from many specialties within daily pharmaceutical practice such as manufacturing, quality control, quality assurance, research and development, engineering, regulatory affairs, marketing, general management and academia (http://www.afiscientifica.it/all/AFI INGLESE.PDF last accessed Aug.4th 2017).

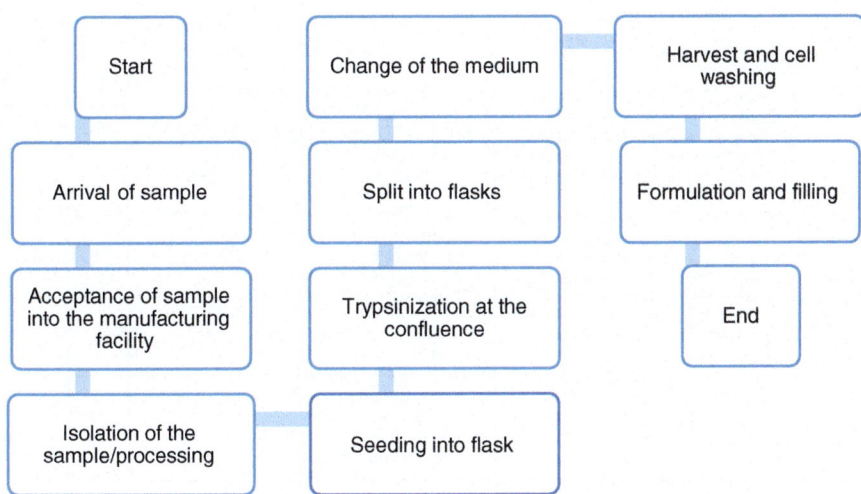

Fig. 6.9 Biotech process subjected to FMEA. Scheme of a standard cellular production including all the relevant steps: incoming procedures, standard process and final formulation for cells growing in adherence

Table 6.4 S, O and D tables used for the FMEA of the biotech process (case study)

Severity		
10	Very serious	Death or significant damage to people
8	Severe	Total damage to the final product
6	Moderately severe	Damage not significant but significant dissatisfaction
4	Moderate	Affects service parameters (quality, time, reliability ...)
2	Medium mild	Slight effect on the parameters
1	Minor	Not detectable by the final user and not affecting the product
Occurrence		
10	Constantly recurring	The error may occur at least once a month
8	Recurring	The error may occur between 5 and 10 times in a year
6	Random	The error may occur between 1 and 5 times in a year
4	Rare	The error may occur once a year
2	Very rare	The error may occur once in 5 years
1	Highly unlikely	The error may occur sometimes in a time of 5–30 years
Detection		
10	**Highly improbable**	**The error can be detected in less than 50% of the cases**
7	Low	The error can be detected in 50% of the cases
4	Moderate	The error can be detected in 70% of the cases
2	High	The error can be detected in more than 90% of the cases
1	Certain	The error can be detected in more than 95% of the cases

Key Process Step or Input	Potential Failure Mode	Potential Failure Effects	S E V	Potential Causes	O C C	Current Controls	D E T	R P N
Formulation and filling	wrong dilution	sample out of specification	10	error in sampling	6	entrusted to the operator's experience	4	240
				manual error by the operator	1			40
	materials not suitable	sample deterioration	8	wrong material handling	1	GMP: SOP	1	8
	wrong filling volume	sample out of specification		instrumental error	2	GMP: SOP	1	20
			10	complexity in aliquot for product characteristics	4	triplicate analysis	4	160

Fig. 6.10 Excerpt from the FMEA of a biotech process. The analysis of the process step considered (formulation and filling) highlighted three possible failures (*wrong dilution, material not suitable* and *wrong filling volume*) and five different causes (from *error in sampling* to *complexity in aliquot for product characteristics*). Among them, only two had a RPN higher than the set threshold (100), because of the highest severity assessment: *wrong dilution* due to *error in sampling* (RPN = 240) and *wrong filling volume* due to *complexity in aliquot for product characteristics (RPN = 180)*

Table 6.5 Improvement actions from the FMEA of a biotech process (case study)

Step	Potential cause	Preventive action	Responsible
Sample isolation/processing/split in flask	Error in traceability of sample/process	Control by a second operator or designated person	Production manager
	Error in performing a critical step	Control by a second operator or designated person	Production manager
First trypsinisation	Operator's mistake (complexity in confluence evaluation)	Control by a second operator or designated person	Production manager
Harvest and cellular washing	Tubes not closed properly	Introduce tube closure check in the batch production record	Production manager
Formulation and filling	Error in sampling	Withdraw a greater number of samples	Production manager/QA
	Complexity in aliquot for special product characteristic	Change formulation for critical products	Production team/QC

Main improvement actions concerned the need for a second operator check, a new check of tube closure, a greater number of samples and a change in the formulation of the product when critical

Fig. 6.11 Summary of risk priority number (RPN) before and after improvement actions. Columns in dark grey represent the number of RPN below the threshold: 24 RPN before the improvement actions and 30 RPN after the improvement actions. The columns in pale grey show the RPNs above the threshold, which are 6 before the improvement actions and 0 after

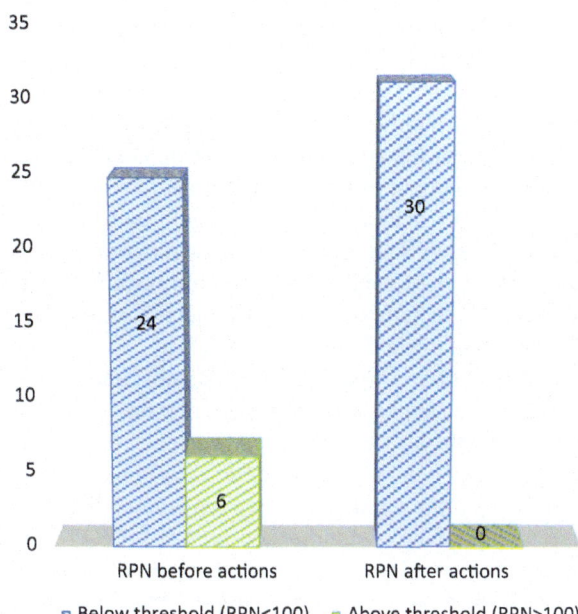

The analysis results revealed that the GMP process was under control; anyway, some improvement opportunities were highlighted. Table 6.5 shows all the suggested actions to reduce the RPN that exceeded the set threshold. These actions are about the need of a control by a second operator or further checks to be inserted in SOP.

This result shows that the risk evaluation defined as severity x occurrence without considering the controls (detection) is not sufficient to completely evaluate the process criticalities. Instead, FMEA, taking into account also the control aspects, provides an additional tool for reducing the chance of a compromised final result.

Another advantage of FMEA is setting up the conditions to maintain under control the process, detecting and evaluating problems. This experience enhances process management, and is fundamental for the design and industrialisation of new processes. In this way, technology transfer from one process to another is done in an analytical and a controlled way.

Among the general benefits that can be outlined, the approach through FMEA allows people to get used to objectively observe the processes.

The application of the methodology required adaptation of the vocabulary to be more appropriate for biotech area. In particular, the term *failure* was changed into *problem* or *error*, more suitable for a biological process.

6.3   Design of Experiment (Antonella Lanati, Andrea Turcato[5])

Statistical experimental design, also known as design of experiments (DoE), is a methodology conceived to plan and conduct experiments that allows to extract the maximum amount of information in the fewest number of experimental runs. DoE was developed in the early 1920s by Sir Ronald Fisher at the Rothamsted Agricultural Field Research Station in London, working on the effects of various fertilisers on different plots of land. The method has been further refined by Yule, Box, Stu and Bill Hunter, Scheffe, Cox, Taguchi and others, so that today it comprises a tool box for virtually any optimisation problem.

One-factor-at-a-time (OFAT) is the most immediate approach to experimentation and is carried out by performing one or more tests for each value (level) of the independent variables (factors), leaving all the other conditions unchanged and repeating the same type of procedure for every single factor. This approach does not take into account the interactions between factors that could be estimated only by varying multiple factors simultaneously. A full OFAT model including all the possible interactions would require a great expense of time and resources as the number of variables increases. A figurative representation of the two different methodological approaches—OFAT and DoE—is illustrated in Fig. 6.12, where the optimal configuration of two factors for the same process is sought. System outputs are portrayed with a greyscale (from pale to dark shades) in a continuous two-dimensional space that is to be explored by experimentation. The four-pointed stars (¤) represent the attempts made by researchers. Every attempt is a combination of two values, one for each factor, and the cardinal number associated with each attempt refers to the experimental step in which it is made. By varying one factor at a time, a greater number of attempts, steps and time is required (1–9¤in Fig. 6.12a). Moreover, the results may suffer from a higher risk of sub-optimality because of the possibility of getting to a relative maximum without gaining a general understanding of the system. In a DoE approach, the first step (marked with 1¤in Fig. 6.12b) identifies four areas, two of which deserve a second experimental step (2¤in Fig. 6.12b), and leads to the identification of the actual optimum. Notably, as we are going to see in the following, DoE also provides an estimate of the interaction among factors that should actually be evaluated with a separate set of experiments with the OFAT method.

6.3.1   The Methodology

To properly design an experiment and plan the extraction of information from the data it generates, it is essential to have a good understanding of the process. A process is fed by inputs and transforms inputs into outputs. When performing a designed experiment, intentional changes are made to the input process variables (called

---

[5]This section is a synthesis of different works by several authors [6, 7].

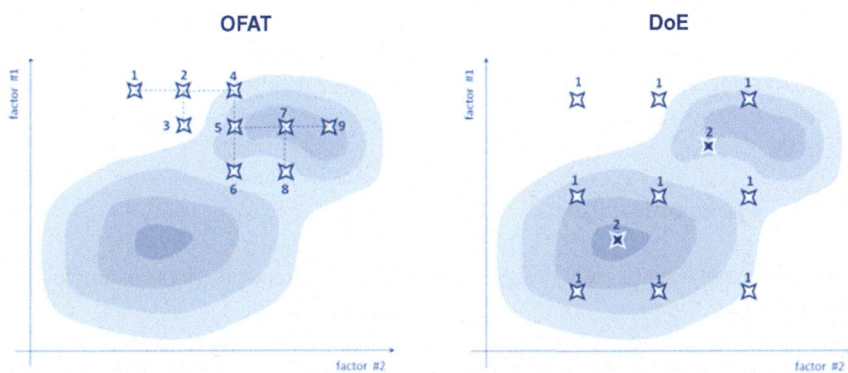

Fig. 6.12 One factor at a time (OFAT) vs. design of experiment (DoE) methodology. Figurative representation of two different approaches to finding the optimal configuration of factors for the same process. System outputs are portrayed with a greyscale (from pale grey to dark grey) in a continuous two-dimensional space that is to be explored by experimentation. The four-pointed stars represent all of the attempts made by researchers. Every attempt is a combination of two values, one for each factor, and the ordinal number associated with each attempt refers to the experimental step in which it is made. (**a**) OFAT: By varying one factor at a time, a greater effort on attempts, steps and time is needed. Moreover, a relative maximum is reached without gaining a general understanding of the system. (**b**) DoE: The first step identifies four areas, two of which deserve a second experimental step leading to the identification of the actual optimum (adapted from [7])

factors) in order to observe corresponding changes in the process *output* or *response*. The different values assumed by each factor are called *levels*, usually for the sake of simplicity varying between two values (namely *high* and *low*) codified by -1 and $+1$, that can be both qualitative and quantitative. Non-controllable variables are responsible for variability in product performance or product performance inconsistency [8].

In order to achieve statistically robust results and relevant conclusions from the designed experiment, it is necessary to adopt different statistical principles that minimise experimental bias that may mask the responses of the significant factors [9]:

- *Randomisation:* Scrambling the running order of treatments assures that more than one factor is changed from a run to the next, avoiding that the pattern of execution and some external conditions (such as changing temperature, material settling, position of samples) could combine to influence the output.
- *Replication*: A correct replication is the duplication of the entire experiment session, or part of it, needed to estimate a possible experimental error. This concept can become clearer when referred to biological and technical replicates: a biological replicate is performed when the same type of organism is grown/treated under the same conditions, while technical replicates are sets of measures performed several times on the same sample.[6] What is needed for a correct application of the replication principle is a biological replicate.

---

[6] For the difference between technical and biological replicates, see [10].

- *Blocking*: It introduces a method to model some known experimental conditions, such as different instruments used during the measurements, different operators performing an experiment or different days of execution, without considering them as factors.

6.3.2   Types of Design

DoE offers a rich set of other designs, to suit most requirements. Some examples are the following:

- Factorial design: The most used set of treatments employed to investigate the effects of factors on a response. It can be *full factorial*, i.e. comprising all combinations of factor levels, or *fractional factorial*, i.e. requiring a reduced set of treatments, upon verification of specific assumptions.
- Plackett-Burman design, which evaluates the effects of main factors only, with a small set of runs.
- Response surface designs (e.g. central composite, Box-Behnken), which are used to identify points of absolute maximum, and to highlight possible non-linearities (for quantitative factors only).
- Mixture design, which is used when factors are components and the output depends on their relative proportions.

As a general approach, a *screening analysis* is first performed with less stringent conditions to identify the most significant factors. For this purpose it is common to work on a large number of potentially significant factors, using designs such as Plackett-Burman or the fractional factorial. Subsequently, an *optimisation analysis* is applied to a narrower set of factors—typically 2–3—to find the best condition that optimises the output(s). On this, selected factors, full factorial, general factorial and response surface (among the others), can be employed to refine the output evaluation, describe possible non-linearities and establish the optimum. This two-step approach results in a significant reduction of experimental runs.

Factorial designs. The simplest factorial design is called *two-level factorial design*, and is used in two forms: *full factorial* and *fractional factorial design*. The two-level factorial design (2^k) uses two levels for all the k variables, and identifies the influence of each and their interaction using 2^k experiments. In Table 6.6 is shown an example of a 2^k design with three factors in randomised order.

Full factorial design generates, with the increasing number of factors, an exponential increase of requested runs. In the case of many factors (e.g. >5), the solution could be a fractional factorial design, which selects only a reduced group of experiments, starting from the most relevant based on specific assumptions, such as ignoring interaction of more than three factors: this is then the *fractional factorial design*.

Table 6.7 helps in understanding the number of experiments required to achieve a robust design, for number of factors of up to 10 and number of runs up to 32. No-framed cell text indicates reduced designs that preserve evaluation of the most useful interactions. For example, in a five-factor design reduced to $2^{K-1} = 16$ runs,

Table 6.6 Example of full factorial design with three factors

Run (standard order)	Run (randomised order)	A	B	C	Output
1	2	+1	+1	+1	Y1
2	3	−1	+1	+1	Y2
3	5	+1	−1	+1	Y3
4	4	−1	−1	+1	Y4
5	6	+1	+1	−1	Y5
6	8	−1	+1	−1	Y6
7	7	+1	−1	−1	Y7
8	1	−1	−1	−1	Y8

First two columns show the standard order attributed to the sequence of experiments (*treatments*) and the run order, randomised to minimise the bias effect on results. Three more columns describe the coded value sequences for the three factors A, B and C, where *+1* refers to the highest level while *−1* to the lowest level. The last column shows the results achieved for each treatment (*output*)

Table 6.7 Guide to the selection of a two-level factorial design

runs \ factors	2	3	4	5	6	7	8	9	10
4	full	III							
8		full	IV	III	III	III			
16			full	V	IV	IV	IV	III	III
32				full	VI	IV	IV	IV	IV

Numbers of factors are listed in columns and the choices for the number of runs are listed in rows. 2^K runs for K factors are full factorial designs and denoted with *full*. For different choices and reduced designs, the resolution degree and the related assessment of reliability are shown: good = *no frame*; acceptable = *dashed frame*; and not acceptable = *double frame*

no main effect or two-factor interaction effect can be confused (*aliased*) with any other main effect or two-factor interaction effect, but two-factor interactions are aliased with three-factor interactions. The ability to distinguish the effects of different factors and interactions in the presence of superposition of effects and interactions (*confounding*) is called *resolution*; most common types of resolution are detailed in Box 6.2. The condition illustrated in the previous example is a *resolution V* and is considered acceptable. Thus this fractional design can adequately estimate two-factor interactions if it can be assumed that three-factor interactions are negligible. Lower resolution designs (dashed frame and double frame in Table 6.7) are affected by the *confounding* of effects, i.e. effects of single factors and of factor interactions are superimposed on the output and consequently indistinguishable.[7]

---

[7] See [11] for further details.

Box 6.2: Main Types of Resolution in a Factorial Design
Resolution III: Main effects are aliased with two-factor interactions, while no main effects are aliased with any other main effect.

Resolution IV: Two-factor interactions are aliased with other two-factor interactions and main effects are aliased with three-factor interactions. No main effects are aliased with any other main effect or two-factor interactions.

Resolution V: Two-factor interactions are aliased with three-factor interactions and main effects are aliased with four-factor interactions. No main effects or two-factor interactions are aliased with any other main effect or two-factor interactions.

6.3.2.1 General Factorial Design

There is also the possibility to consider more than two levels for each factor, and to create a *general factorial design* [11]. The number of runs requested by a general factorial design dramatically increases with the number of levels for each factor and the number of factors. For example, while a full factorial design on three factors at two levels requires $2^3 = 8$ runs, a general factorial design with three factors, one of which at three levels, requires $2^2 \times 3 = 12$ runs, but if the factors changing on three levels are 2, then the number of runs increases to $2 \times 3^2 = 18$. For a five-factor design, only one of which at three levels, the number of runs needed is $2^4 \times 3 = 48$. Moreover, according to Anderson and Whitcomb [11], a general factorial design is to be avoided when the number of factors increases (typically higher than 3), and even the reduction of a general factorial design requires heavy, *ad hoc* elaboration. The same authors suggest making preliminary tests to attempt to reduce the analysis to a two-level factorial. Furthermore, the normal probability plot for effects—one of the statistical outputs of the DoE analysis, supplying the statistical significance of variations—cannot be computed for general factorial design.[8] Therefore, most commercial software programmes supporting the DoE do not provide this plot, while other software products use workarounds. This limitation makes it more difficult to understand immediately which factors are statistically significant. Globally, general factorial is mainly used in refining analysis (optimisation, see above).

6.3.2.2 Response Surface Design

Response surface designs (e.g. central composite, Box-Behnken) are used to refine the design in the phase of response optimisation, identifying points of absolute maximum and highlighting possible non-linearities (for quantitative factors only). The mathematical model is enriched with quadratic (second order) terms that can

---

[8] The *normal probability plot* is generated from the coefficients of the regression model that for the general factorial design is substituted by an evaluation of means of levels.

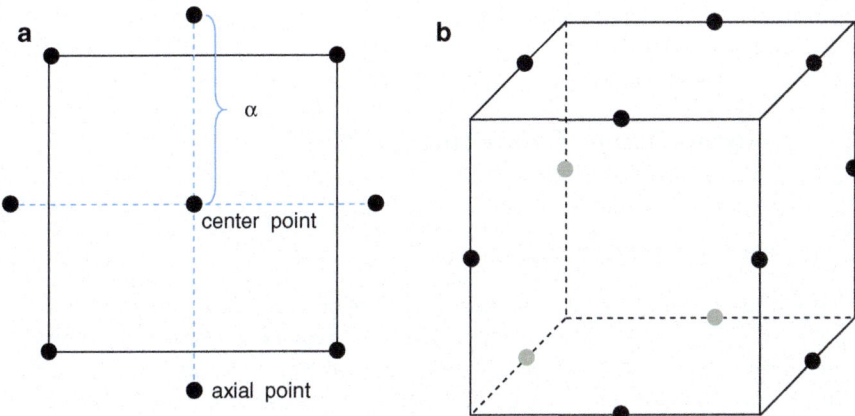

Fig. 6.13 Central composite and Box-Behnken designs. (**a**) Experimental points for two factors x_1 and x_2 in a central composite design: the vertices of the square are the experimental points of a factorial design and in this case the axial/star points lie outside the square ($\alpha > 1$). (**b**) Box-Behnken design for three factors. It includes the middle points of the edges and does not include the vertices of the cube which are the experimental points of the factorial design

account for curvatures in the response. This is achieved adding points to the factor space, according to two main types of response surface design: *central composite design* and *Box-Behnken design*. Both require continuous factors, i.e. not assessed by means of a qualitative scale.

Central composite design is the most commonly used response surface design and can be carried out by simply adding *centre* and *axial points* to a previous factorial experiment. Figure 6.13a) illustrates the position of the experimental points for two factors x_1 and x_2, according to this response surface design. The experimental points of the full factorial design for factors (x_1, x_2) lie on the vertex of the square. The point in the middle of the square is called *centre point*. Points outside the square are called *axial* or *star points* and in this example are distant from the central point by a quantity α greater than 1 (*circumscribed* design—CCC). The parameter α may also be equal to 1 (*face-centred* design—CCF), as well as less than 1 moving the axial points inside the square (*inscribed* design—CCI). The choice among the three options is made considering the possibility of using points outside the original experimental space and the precision of the response prediction—the greater the experimental space, the better the prediction.

Box-Behnken design does not contain a factorial design. All experimental points are the midpoints on the edges of experimental space, for example the square for two factors or the cube for three factors. Figure 6.13b) illustrates the experimental configuration for a three-factor design. Box-Behnken design offers some advantage in terms of runs for number of factors up to 4, but is less precise in some areas missing the corners of the experimental space. However, it may be useful if the application does not include extreme values for the factors.

Response surface designs are used to estimate and optimise the so-called *design space* of the parameters of a pharmaceutical process, in order to achieve more flexibility in the marketing authorisation for a new drug.[9]

6.3.2.3 Mixture Design (Franco Pattarino[10])

In many situations, the object of the study is a mixture, i.e. a system that is formed by two or more components (x_i) that are interdependent. In other words, the proportions of the mixture components add up to 1 ($\sum_q^{i=1} x_i = 1$).

Because of this constraint, the geometric description of the factor space for a mixture is different from that of the independent factors involved in non-mixture system. The domain of existence of a mixture is called simplex and its form can be a line (two-component mixture), a triangle (three-component mixture) or a pyramid (four-component mixture) (Fig. 6.14): each point on or inside the simplex represents a mixture, identified by a coordinate system.

Particular (i.e. specific) experimental designs, different from those employed for systems of unconstrained factors, are required in studying mixture systems: as an example, for a mixture consisting of three components, the simplest form of experimental design is illustrated in Fig. 6.15a).

The experimental setting of this design consists of nine points, all located on the boundary of the simplex diagram, corresponding to all possible combinations of the

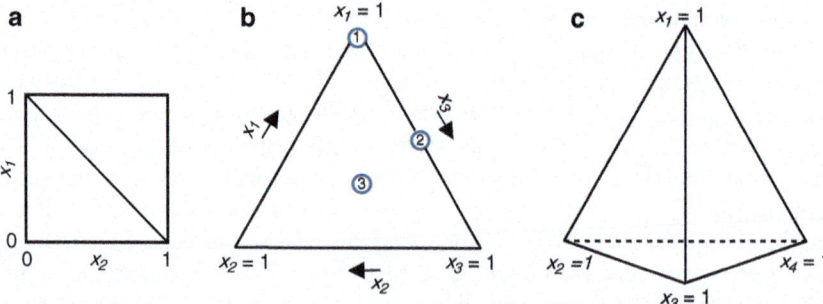

Fig. 6.14 Simplex diagram for mixtures with (**a**) $q = 2$, (**b**) $q = 3$ and (**c**) $q = 4$ components. The vertices represent mixtures composed by one component only; for example, the point 1 in (**b**) is the mixture consisting of the x_1 component only with coordinates (1, 0, 0). The points lying on the edges correspond to mixtures where one of the components is absent; for example, in (**b**) point 2 on the left edge of the triangle represents the binary mixture with coordinates (1/2,1/2, 0). Each location inside the simplex represents a different blend of all the components of the mixture (e.g.: point 3 in (**b**) is a three-component mixture with coordinates (1/3, 1/3, 1/3))

---

[9] See [12] for further details.

[10] Professor of Pharmaceutical Technology, University of Piemonte Orientale, Italy.

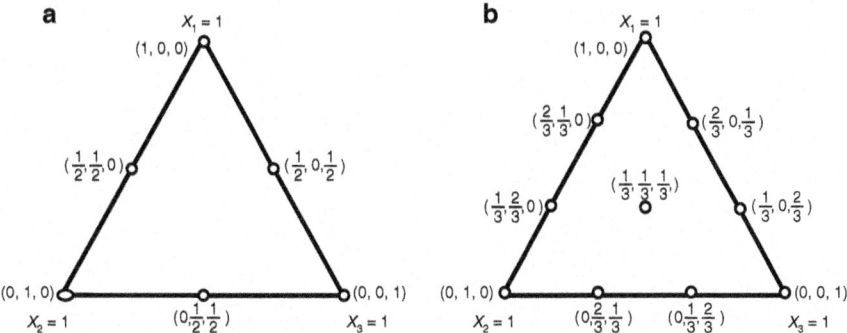

Fig. 6.15 Experimental designs for a ternary mixture ($q = 3$). (**a**) The simplex lattice design consisting of six experiments, located at the vertices (mono-component) and in the centre of each edge (binary-component mixtures); (**b**) the simplex centroid design requiring ten experiments, three located at the vertices (mono-component), six on the edges (binary component) and one at the centre of the simplex (three-component mixture)

three levels (0, ½, 1) of each component. Other designs are possible that consist of a higher number of experimental points derived from the combination of the mixture's components taken in different proportions. In Fig. 6.15b) the setting of a simplex centroid design for a three-component mixture is reported.

The properties of a mixture are strictly related to the proportions of each of the system's components and a special correspondence exists between the design setting and the mathematical equation to be fitted to the experimental response values. The response data from an experimental mixture design can be fitted to a model whose degree corresponds to the number of mixture components and that contains a number of terms equal to the experimental points of the design. For the simplex lattice design of Fig. 6.15a), the model to be used is the quadratic polynomial:

$$Y = \sum_{q}^{i=1} \beta_i x_i + \sum \sum_{q}^{i=1} \beta_{ij} x_i x_j$$

It should be noticed that this polynomial does not contain the β_0 parameter (present in the models employed in data analysis coming from factorial designs). The parameter β_i represents the expected response for the pure component x_i and the $\sum \beta_i$ is called the linear blending portion. The parameters of higher degree (β_{ij}) account for the curvature of the response surface and are named synergistic or antagonistic blendings.

The fitting of data to a model (including an appropriate number of replicates for error evaluation) can be carried out using the ordinary least square method with a software programme able to provide the model parameters and statistics for the experimental models of the mixture.

6.3.3   Analysing Phase

Once the experimental plan has been defined and the experiments executed as designed, the corresponding output values are evaluated to seek out the most influencing factor or combination of factors. DoE quantifies this influence on the output by calculating the **effect** of each factor or combination of factors as

$$\text{Effect}_y = \frac{\sum Y_+}{n_+} - \frac{\sum Y_-}{n_-}$$

where Y_+ and Y_- are the output measures for, respectively, the higher and the lower level of the factor (or interaction) y, and n is the number of instances.

After performing experiments according to the planned design, results are analysed through graphical interpretation and a set of statistical parameters.

Statistical Plots: Normal Probability Plot and Pareto Plot. The *normal probability plot* is a different representation of a distribution, with the cumulative percentage on the logarithmic Y-axis and the ordered values of the observations on the X-axis. In this representation, the Gaussian distribution appears as a straight line. It is used to check normality of the data and to find out the most significant ones: non-significant data are dispersed along a straight line, whereas significant data are away from it. In the experimental design, the normal probability plot is used to evaluate significance and normality of both main and interaction effects [11].

The *Pareto plot* displays the absolute values of main and interaction effects: a reference line shows statistically significant values (usually drawn for $P < 0.05$).

Factorial Plots: Main Effect, Interaction and Cube Plot. A *main effect plot* is the plot of the mean response values at each level of a factor; it is used to compare the relative strength of the effects of various variables on the output. It appears as a straight line, where the slope indicates the direction while its magnitude the strength of the effect.

An *interaction plot* shows how different combinations of factor settings affect the response: non-parallel lines show interaction between pairs of factors [8].

Cube plots display the average response values at all combinations of process or design parameter settings. This allows to easily determine the best and the worst combinations of factor levels to achieve the desired optimum response.

Contour Plot and Surface Plot. Both *contour* and *surface plots* are used to portray the relationship between the response and two continuous factors. In DoE they are mainly employed to represent the results of a response surface design. Contour plot is a two-dimensional diagram in which points that have the same value are connected to draw the so-called *contour lines* or *isolines*. Sometimes the areas between the isolines are coloured with decreasing intensities, as illustrated in Fig. 6.16a). The same information can be shown in a three-dimensional representation with the *surface plot* (Fig. 6.16b), where the two factors are displayed on the x-axis and y-axis, and the response z is described with a surface function of x and y variables in the Cartesian space.

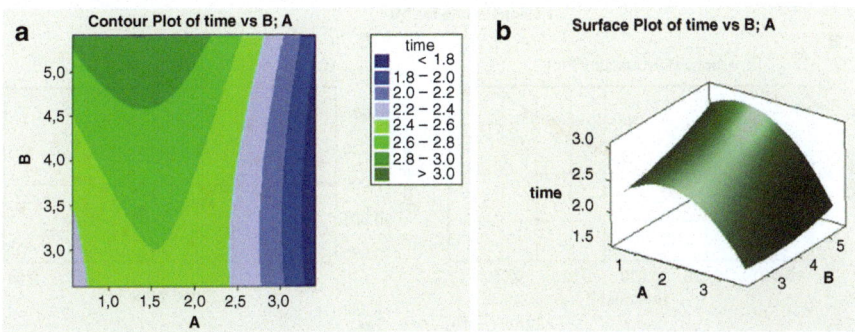

Fig. 6.16 Contour and surface plots. (**a**) Contour plot of a two-dimensional experimental space: The response *time* depends on the two factors *A* and *B*. Points that have the same value are connected to draw *contour lines* or *isolines*. The areas between the isolines are coloured with decreasing intensities. (**b**) Surface plot of the same experimental space. The relationship between the response *time* and the two factors *A* and *B* is represented by a surface function in a three-dimensional Cartesian space

Contour and surface plots show the non-linearities of the response in the experimental space. They also allow locating the optimum response area and addressing further refinements of the analysis.

Regression Model. The DoE also provides a mathematical (regression) model illustrating the relationship between the response and the set of design factors. The generic equation of the following is eq.1

$$y = \beta_0 + \beta_1 x_1 + \beta_2 x_2 + \beta_n x_n + \beta_{1,2} x_{1,2} + \ldots + \varepsilon$$

where $x_1, x_2 \ldots x_n$ are the factors and $x_{1,2}, x_{2,3} \ldots$ are their interactions, β_i and β_{ij} are the regression coefficients provided by the method and the term ε is the random error. The mathematical meaning of coefficients in a multiple regression is the weighting that every effect assumes: the higher the absolute value of the coefficient, the larger the effect a specific variable has on the output. The sign of each term indicates the positive or negative effect of the factor or factor interaction on the output. The random error should be distributed approximately normally and independently with a mean of zero and constant variance (this last condition is call *whiteness*). This regression model equation can be used to predict the response for different combinations of variables at other levels, if lying within the range identified by the levels chosen for the analysis. The response values obtained from the previous equation are called *predicted values* and the actual response values obtained from the experiment are called *observed values*.

Residual Analysis. The appropriateness of the mathematical model can be effectively visualised by means of simple plots that show residuals that are the difference between observed and predicted values. Residuals must be low in magnitude and distributed normally. Moreover, the DoE analysis is based on statistical methods (ANOVA and hypothesis test), which assume the normality of the data; it must be checked that the residuals are normally distributed, because their distribution is a qualitative

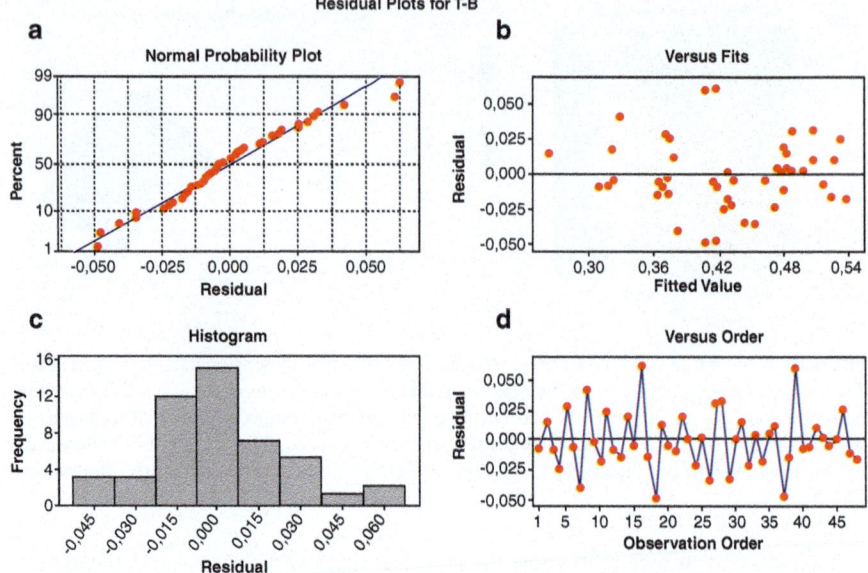

Fig. 6.17 Residual plots. A standard representation of information regarding residuals includes (**a**) a *normal probability plot* and (**b**) a *histogram*, both allowing to check for normal distribution of data. (**c**) *Versus fit* plot is used to check for constant variance and (**d**) *versus order* plot allows verification of the assumption that the residuals are uncorrelated to each other

parameter describing the correctness of the underlying statistical model. Different plots are available for this check: the *histogram of residuals*, the graph that correlates the residuals with their frequency, must appear approximately symmetric and bell shaped, to confirm the normal distribution of residuals. The *versus order plot* must show randomly scattered residuals with the absence of significant patterns in the distribution, to demonstrate the time independence of residuals. The *normal plot for residuals* is conceptually the same as the one used for effects and interactions: residuals must lie on a straight line if the data have a Gaussian distribution. To confirm the assumption of a constant variance, the *versus fit plot for residuals* must not show abnormal patterns (e.g. a loudspeaker shape). Figure 6.17 shows examples of residual plots.

To deal with a non-normal distribution of data, variance-stabilising transformation can be applied on data, e.g. calculate the logarithm; for detailed understanding of this analysis, refer to Box-Cox methods, described in more specific studies.[11]

6.3.4   Example of Application: A Simple Full Factorial Design

The optimisation of the configuration of an assay based on the electrical stimulation of cells and the optical recording of membrane potential changes can provide a simple example to understand how to plan and execute a full factorial design. Let's imagine

---

[11] For example, see [9].

that the assay is performed stimulating with a square pulse of current cells that are previously loaded with a voltage-sensitive dye. The fluorescence intensity changes, due to the electrical stimulus, are recorded by the microscope camera and subsequently elaborated. An experimental design can be planned in order to obtain the best cell response, taking into account several factors that may influence the output:

- Concentration of the dye: The higher the fluorescence, the better the signal-to-noise ratio but also the potential toxic effect of the dye.
- Intensity of the current pulse: The higher the intensity, the higher the cell response but also the possibility of membrane electroporation.
- Camera gain: The higher the gain, the stronger the fluorescence signal recorded but also the noise.

The *response breadth [%]*, as the percentage of variation due to the current stimulus with respect to the nominal value, is chosen as the output.

Each factor is set on two levels. This design with three factors, each on two levels, needs $2^3 = 8$ treatments, which become 16 with 2 replications. The runs are randomised and the *response breadth [%]* output recorded after the execution of each treatment (Table 6.8). There is no need to reduce the design, because 16 runs are clearly affordable for this kind of assay.

The first analysis should be devoted to residuals, which are approximately normally distributed (Fig. 6.18).

Table 6.8 Full factorial design for the cell stimulation example, with three factors, two levels, two replicates

StdOrder	RunOrder	Gain	Current	Dye	Response breadth [%]
1	6	−1	−1	−1	0.84
2	1	1	−1	−1	15.49
3	3	−1	1	−1	21.01
4	8	1	1	−1	28.74
5	14	−1	−1	1	12.03
6	11	1	−1	1	17.79
7	15	−1	1	1	17.90
8	16	1	1	1	4.89
9	5	−1	−1	−1	4.25
10	10	1	−1	−1	10.80
11	9	−1	1	−1	28.14
12	2	1	1	−1	27.74
13	13	−1	−1	1	16.19
14	7	1	−1	1	7.76
15	4	−1	1	1	15.81
16	12	1	1	1	15.56

The treatments are numbered following a standard order (StdOrder, 1 to 16), but executed in a randomised order (RunOrder). Levels are coded as −1 and +1. Each treatment is duplicated: first 8 rows are identical to second 8 rows. The last column shows the output obtained from the execution of treatments

Fig. 6.18 Residuals for the cell stimulation example. (**a**) Histogram shows a normal distribution. (**b**) Versus fit plot shows no abnormal distribution of data and confirms a constant variance

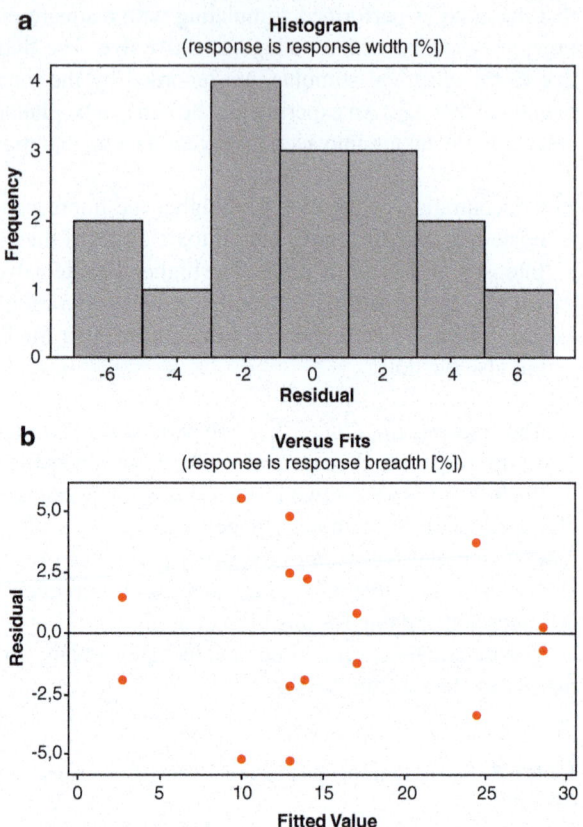

The Pareto plot and the normal plot (Fig. 6.19) show the same information, i.e. the factors and relative interactions ranked by statistical significance. The current (factor B in the graphs) is the most influencing factor, together with its interaction with the dye (BC). This latter is also influencing the output but only when interacting with the camera gain (AC): this appears to be meaningful, as both gain and dye influence the fluorescence response. On the other hand, also current and dye are likely to contribute to a larger response. Other factors and interactions don't reach the significance threshold: they are under the red line in the Pareto plot and not labelled in the normal plot.

Now the analysis of the main effect plot and the interaction plot can be enlightened by the above considerations: the only behaviours that matter are those of current in the main effect plot (Fig. 6.20), as well as Current+Dye and Gain+Dye in the interaction plot (Fig. 6.21).

Of note, the slope of the current in the main factor plot is the largest, confirming that current is the most influencing factor. Best results are obtained with the larger value (coded level *1*).

In the interaction plot, at the intersections of current and dye (graph in dotted frame), it can be observed that the two slopes for current = −1 and current = +1 are significantly different. The same observation can be done for the interaction between

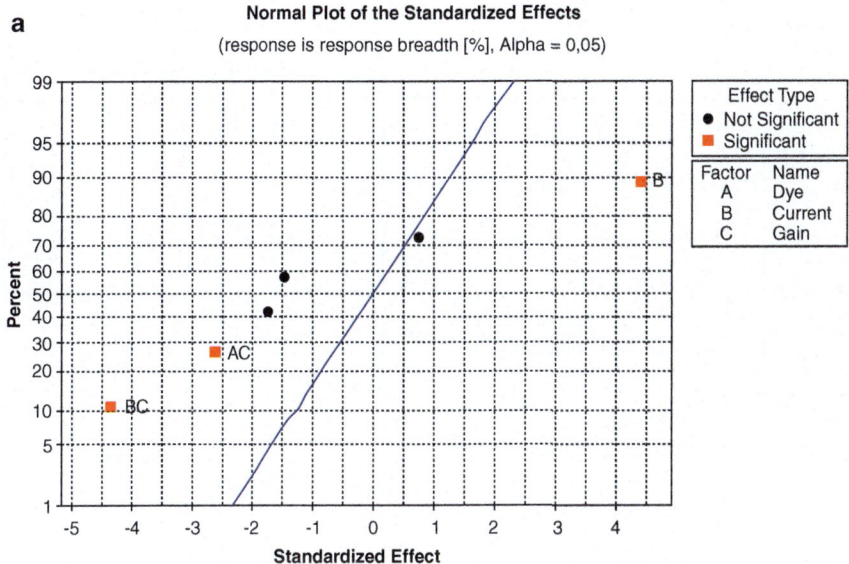

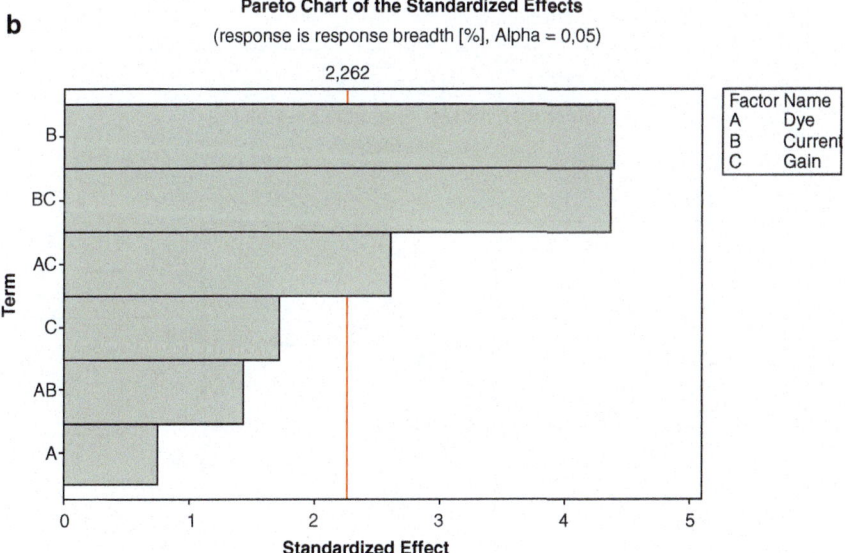

Fig. 6.19 Statistical plots for the cell stimulation example. Both plots confirm that the significant factors and interactions are current (B), current + dye (BC) and dye + gain (AC). (**a**) Normal plot: significant factors are marked with a label. (**b**) Pareto plot: significant factors are above the threshold (vertical line)

dye and gain (graph in dashed frame). Finally, the cube plot shows the best combination of factors leading to the widest response (Fig. 6.22): the three dimensions are, respectively, gain (X-axis), current (Y-axis) and gain (Z-axis). Each vertex is labelled with the output value corresponding to the combination of values of the three factors.

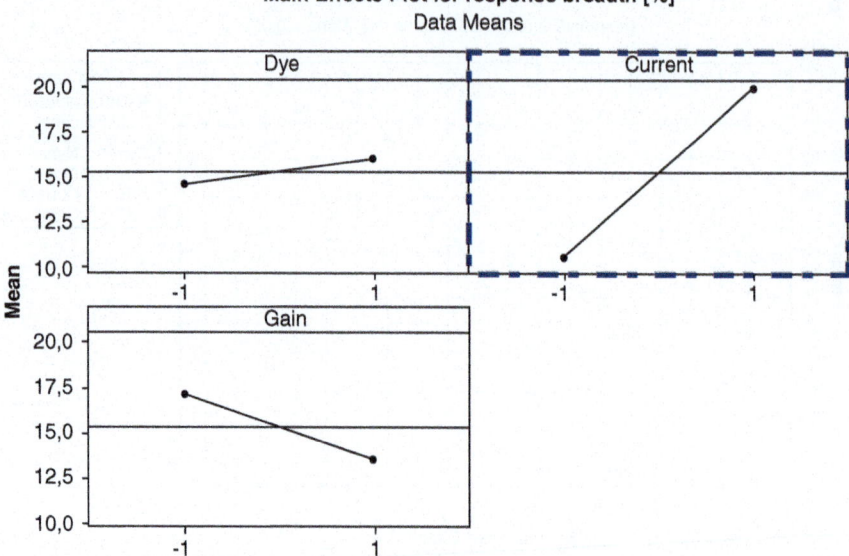

Fig. 6.20 Main effect plot for the cell stimulation example. The only significant factor is current (*dashed frame*): the narrowest effect on *response width* is obtained for the lowest Current value (−1) while the best *response width* is obtained with the highest current value (+1)

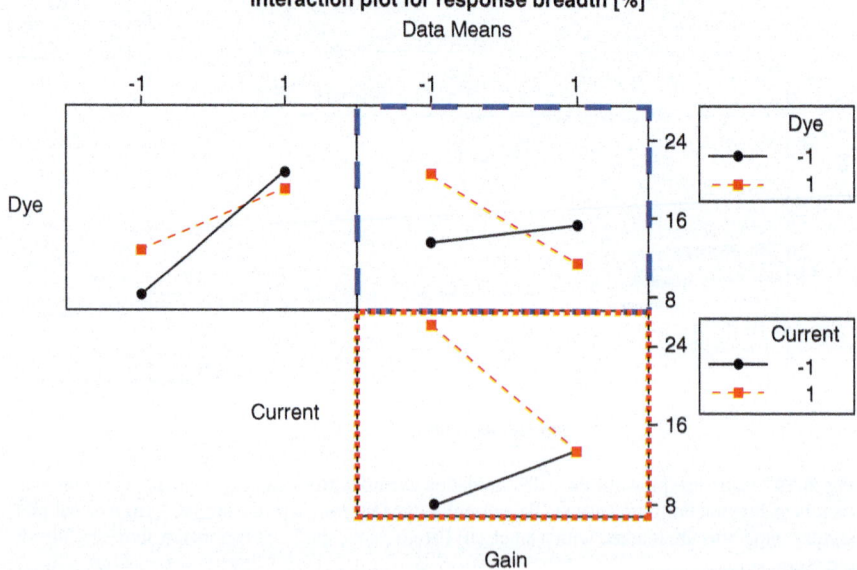

Fig. 6.21 Interaction plot for the cell stimulation example. The two significant interactions are Current + dye (*dotted frame*) and dye + gain (*dashed frame*). In the first one, effects (values referred on the right axis) show a significantly different slope between the *dashed line* for gain = +1, and the *solid line* for gain = −1. A similar behaviour can be noted for dye + gain

Fig. 6.22 Cube plot for the cell stimulation example. It shows the relationship between the three factors and the *response breadth*. The three dimensions of the cube correspond to the three factors. The value on the vertex, circled with a *solid line*, is the best response that can be achieved with the chosen factor values

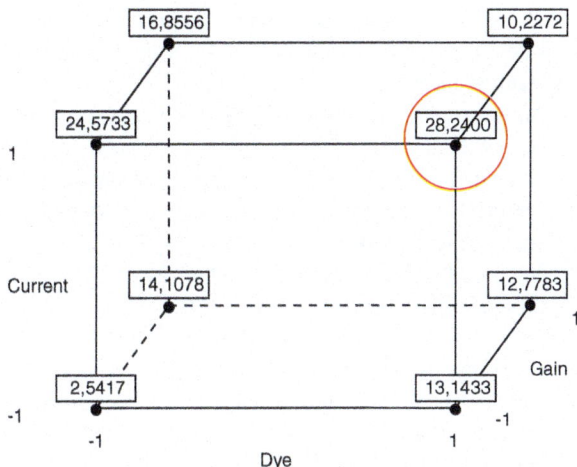

Cube Plot (data means) for response breadth [%]

As it can be seen, the vertex circled in red corresponds to the maximal response breadth and refers to the triplet gain = 1 (max); current = 1 (max) and dye concentration = −1 (lowest), which optimises the assay response for the considered range of values.

6.3.5 Example of Application: Optimising an Image Analysis Protocol for Drug Screening [6]

An unconventional DoE approach has been followed for optimising the parameter configuration of a software programme developed in-house for a research project. At Alembic, the bio-imaging facility of the San Raffaele Scientific Institute, a multidisciplinary team of researchers developed an innovative optical platform for ion channel drug screening, based on a proprietary approach. Briefly, cells expressing the ion channel of interest are loaded with a fluorescent voltage-sensitive dye, and the effect of a drug is then revealed by the fluorescence values recorded before and during exposure to electrical stimulation.[12] To analyse the images captured by the system, a software programme was developed in-house, named MaLIA (MatLab Image Analysis). The MaLIA offers the possibility to employ different filters, parameters and types of analysis. Data representative of cellular response to the electric pulses are used to extrapolate changes in resistance/conductance; these values are put in relation to increasing concentrations of the molecule of interest, to obtain a typical sigmoidal concentration-response (CR) curve. In the course of the project, the MaLIA evolved, gaining in flexibility allowing to explore multiple analytical options.

---

[12] See EP2457088 patent for more details.

In the development phase, different analysis configurations were trialled: the parameter space was narrowed to a set of five, four of which varying between two values only. DoE was employed to define the optimised values of these parameters, in order to perform a standard analysis in full automation, without external, arbitrary interventions.

An unusual approach was trialled: while a standard screening phase is intended to narrow the number of factors, this study needed to reduce the number of levels of just one factor, to perform an optimisation on the whole set of parameters. To do this, some hypotheses were tested, and then confirmed within the validation phase for the obtained results.

For the DoE analysis, the following five software parameters (factors) appeared to influence the output data:

(a) Binning, used to reduce image noise by combining clusters of pixels into single pixels: It can have the values 1, 2 or 4, determining the effectiveness of the filtering action.
(b) Shape mask (ShapeM), used to select the membrane-responsive areas in the cell image: It can be enabled or disabled.
(c) Minimum responses filtering (MinRespFilt), used to discard signal values lying inside the noise range: It can be enabled or disabled.
(d) Response calculation (RespCalc) may be chosen between fold change (FC) and normalized fold change (NFC).
(e) Output data filtering (OutputDataFilt) can be purely statistical or functional, i.e. excluding variations not coherent with the expected biological response.

Two outputs were also identified to evaluate the influence of these parameters on CR curves: the R-squared (Rsq), as a measure of good fit of the sigmoidal curve, and *top minus bottom* (T-B), as the difference between highest and lowest values in the sigmoidal curve (a measure of the efficacy of tested drug). Finally, in order to account for possible inter-day variations (due to biological variability and/or changes in the process), the researchers repeated the same set of treatments on image stacks obtained on three different experimental days, and modelled each of these replications by blocks.

A first *screening* phase was planned with a *general full factorial design* on the binning and the sole factor (ShapeM) able to interact significantly with the binning, being the only two directly related to the pixels of the image. This assumption was then verified on the final result.

The general full factorial design (18 runs, 3 replicates) clearly demonstrated an interaction between binning and shape mask on the Rsq output but not on the T-B output. Moreover, the analysis showed significantly lower values for both Rsq and T-B when using binning 1 compared with binning 2 and even lower values with binning 4. A P-value $= 0.248$ for the variable blocks showed no influence of inter-day conditions on Rsq, while a P-value < 0.001 indicated a significant influence on T-B.

Following these results, the level $= 1$ for binning was discarded and a factorial analysis could be performed on the five factors, all on two levels. The researchers, interested also on the influence of inter-day variability, decided to produce three replicates running MaLIA on image stacks produced in different days. A full factorial design would have required 32 runs per replicate, i.e. a total of 96 runs. In order to reduce this number, they assumed that the interactions of the second order were sufficient for a correct approximation. The number of trials was thus reduced by employing the fractional factorial design with resolution V, which required 48 runs for 3 replicates. Figure 6.23 shows the results for Rsq and Fig. 6.24 for T-B.

These results allowed the researchers to identify one optimised set of parameter values, for both Rsq and T-B.

Some validations were then performed, first to confirm that the binning was solely interacting with ShapeM and that its discarded value of 1 really generated the worst results on Rsq and T-B. It was the proof that only one interaction between two factors was significant and consequently it was justifiable to ignore third-order interactions. The third validation aimed at verifying that the obtained optimal set of parameters actually produced best results on Rsq and T-B. Then MaLIA was run on different image stacks, obtained on different experimental days and with different cell lines and drugs. The tests confirmed that binning = 1 always produced worse results compared with binning = 2 or 4. Moreover, with only one exception in ten

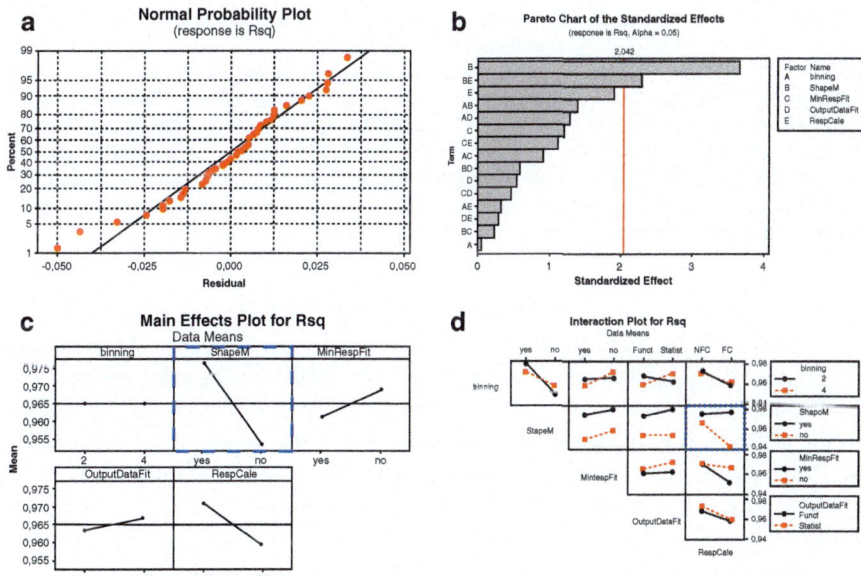

Fig. 6.23 Results of the fractional factorial design (Res. V) for Rsq. (**a**) Plot for residuals indicates a suitable distribution of residuals. (**b**) The Pareto plot of the standardised effects indicates that there are only two significant effects (i.e. lying beyond the vertical line that marks the threshold for alpha = 0.05): shape mask and the interaction between shape mask and response calculation. (**c**) The main effect plot confirms that the shape mask effect (*dashed frame*) is the most important among the single factors and that best results are obtained when the mask is applied. (**d**) The interaction plot shows the best combination for shape mask and response calculation (*dotted frame*): if ShapeM = yes, both values for RespCalc are suitable

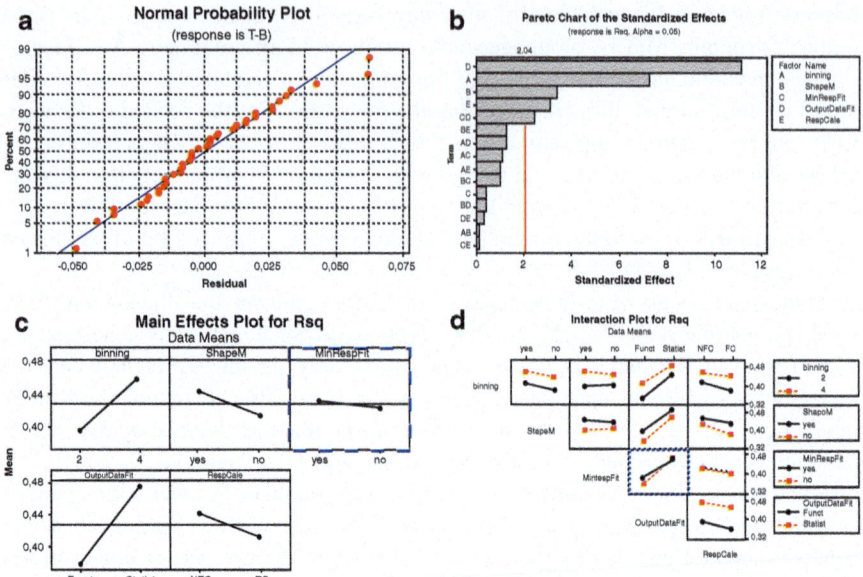

Fig. 6.24 Results of the fractional factorial design (Res. V) for T-B. (**a**) The normal probability plot for residuals indicates a suitable distribution of residuals. (**b**) The Pareto plot of the standardised effects shows that all single factors, with exception of the minimum response filter (MinRespFilt), are significant, while only one interaction, the one between MinRespFilt and OutputDataFilt, lies beyond the vertical line (threshold for alpha = 0.05). (**c**) The main effect plot indicates the best values for the significant factors: binning = 4, ShapeM = yes, OutputDataFilt = Statist and RespCalc = NFC. MinRespFilt (*dashed frame*) is not significant. (**d**) In the interaction plot the value MinRespFilt = no, together with OutputDataFilt = Statist (*dotted frame*), is the value of the significant interacting factors that optimise the output T-B

experiments, the optimised parameter set produced the best C-R curve in terms of Rsq and T-B. Figure 6.25 shows a comparison between a CR curve obtained with the optimised set and a suboptimal one (namely the worst) on the same image stack.

The influence of inter-day conditions revealed by the use of blocks was attributed to the different conditions of the equipment during the development of the drug screening platform and was further investigated before the end of the project.

As a final consideration, this unconventional approach can be verified with respect to saving of number of runs and consequently effort. Let us consider only one replication. With an OFAT approach, the researchers had to run 48 ($2^4 \times 3$) experiments, but missed information on the interactions. With a general factorial on five factors, one of which at three levels, the runs were to be the same number but would not allow to easily obtain information about the statistical significance of the effects. With a full factorial design on five factors at two levels, they would have run 32 (2^5) experiments, having however discarded binning = 1 without checking for justification. With this novel design, the researchers performed 6 (2×3) runs for the first phase, determining the best values for binning, and 16 ($2^{(5-1)}$) for the second

Fig. 6.25 Concentration-response curves obtained on the same image stack with the optimised and a suboptimal parameter set. The CR curve shows the fractional changes of the membrane resistance at different drug concentrations (log). The CR curves obtained with (**a**) the suboptimal and with (**b**) the optimised parameter sets on the same image stack are compared to highlight the marked improvement: Rsq from 0.91 to 0.99, T-B from 0.28 to 0.49

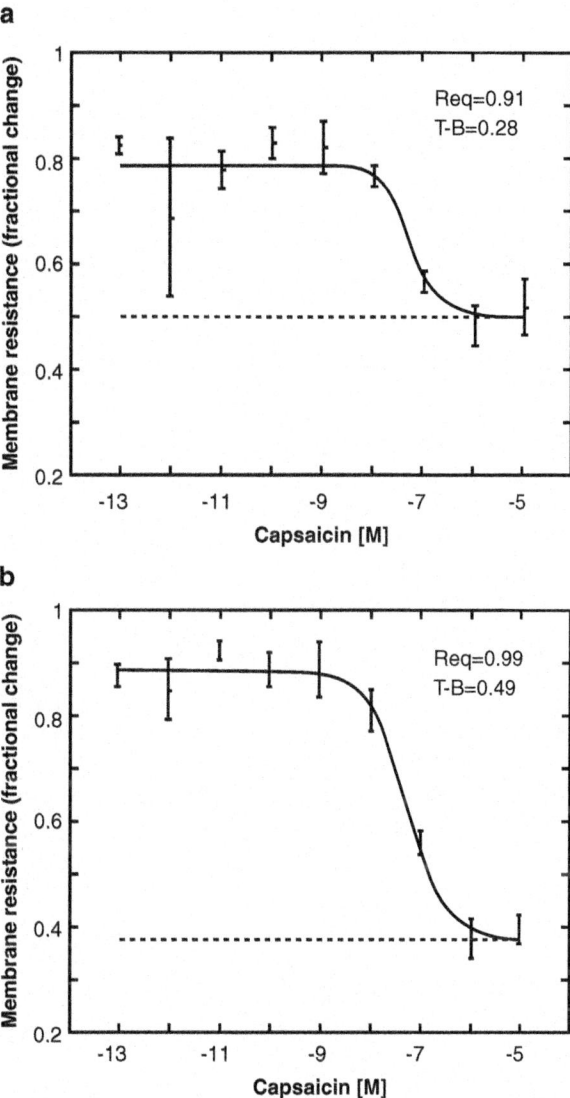

phase by just ignoring third-order interactions, for a total of 22 runs for each replicate. Each MaLIA run (inputting data, setting parameters, waiting for analysis elaboration and collecting results) took a skilled operator, at least 7–10 min. Accordingly, the researchers saved up to 13 h on a total predicted effort of 24 h, i.e. ~54% saving. Overall, by this DoE approach, they saved effort, at the same time gaining more information. It is worth noting how a conscious introduction of constraints to reduce the degrees of interactions, together with a two-stage design, can greatly simplify both the modelling and the achievement of results.

6.3.6   Example of Application: Mixture Design (Franco Pattarino[13])

A mixture experimental design was used for studying the milling of a solid substance (x_1), carried out in the presence of a surface-active agent (x_2) and of a polymer (x_3): the design setting is reported in Table 6.9 together with the logarithmic values of the response, i.e. particle size of component x_1, after comminution.

The experimental data were fitted to a seven-term polynomial equation (cubic model), resulting in

$$y = 6.725\,x_1 + 7.949x_2 + 6.033x_3 - 3.349x_1x_2 - 1.567x_1x_3 - 2.747x_2x_3 + 0.808x_1x_2x_3$$

The coefficients of the linear terms are in the rank $\beta_2 < \beta_1 < \beta_3$ and this leads to conclude that component 2 produces particles with the greatest dimensions. Moreover, all the quadratic terms are negative: each binary blend of the three components producing particles with smaller size than that expected by averaging the response values of the pure blends.

The results of the modelling can also be analysed and interpreted using a graphical representation. In Fig. 6.26 the contour plot of $ln(Particle\ size)$ estimated by the model is illustrated: upon close examination, it can be seen that the lowest logarithmic value for particle size would be obtained with a mixture composed of 44% x_1, 14% x_2 and 42% x_3.

Table 6.9 Design setting in the milling of the three-component mixture

Mix no.	x_1	x_2	x_3	Particle size (nm)	Ln(Particle size)
1	1	0	0	873.0, 794.7	6.772, 6.678
2	0	1	0	2790.0, 2875.6	7.934, 7.964
3	0	0	1	436.3, 398.2	6.078, 5.987
4	$\frac{1}{2}$	$\frac{1}{2}$	0	650.8, 679.3	6.478, 6.521
5	$\frac{1}{2}$	0	$\frac{1}{2}$	393.5, 403.0	5.975, 5.999
6	0	$\frac{1}{2}$	$\frac{1}{2}$	556.1, 537.5	6.321, 6.287
7	$\frac{1}{3}$	$\frac{1}{3}$	$\frac{1}{3}$	417.5, 412.0	6.034, 6.021

Each factor is set at the lowest (0), medium (1/2) and highest (1) level. As illustrated in Fig. 6.15a), the needed runs are 6, plus a midpoint at 1/3 of the range of each factor

---

[13] Professor of Pharmaceutical Technology, University of Piemonte Orientale, Italy.

Fig. 6.26 Contour plot of *ln(Particle size)* as estimated by the model. In a contour plot, points on the contour lines (isolines) represent mixtures that have the same response value. In this example, the isolines for *ln(Particle size)* show worse responses as x_2 increases. The lowest particle size is observed for the mixture constituted by 44% x_1, 14% x_2 and 42% x_3

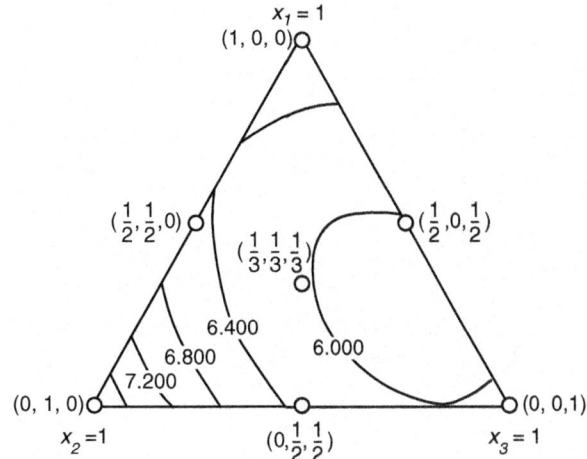

6.4 Lean Management and Six Sigma

Lean management and six sigma are two methodologies developed in the manufacturing industry (in Toyota and Motorola, respectively), which in the last decades have been translated to different fields, such as the service industry, health, pharma, public administration,[14] analytical laboratories and even clinical and translational research [14]. Both methodologies, born from the TQM culture, share the same attention to the customer, effectiveness in execution and quality of results while offering different techniques.

The main goal of the lean approach is to eliminate waste and is achieved by means of various methods and tools, with the common view of improving the value perceived by the customer. The six sigma approach, driven by the voice of the customer (VOC), concentrates on the optimisation of production processes, reducing opportunities for error by means of statistical tools. Both make extensive use of continuous improvement plans. Seeing the advantages that these two disciplines can provide to scientific research may not be so intuitive. After the discussion of their main concepts, a paragraph will delve into their applicability to research.

6.4.1 Lean Management

At its origin, this methodology was called Toyota Production System (TPS) and was developed by Taiichi Ohno, an engineer working at the car manufacturing plant. During his career, he developed a philosophy and devised several methods to eliminate waste throughout the production line. Once introduced in the Western world

---

[14]As an example, the application of Lean Thinking in the Washington State Government offices, see [13].

TPS was (somewhat) restructured and integrated into lean production. Subsequently, with its diffusion in fields other than manufacturing, it was called lean management, more recently lean thinking. Beyond all these epithets, the essence of the method remains the same: reduce waste, increase productivity, increase quality, strengthen the organisation and expand the market. Lean management is based on the *4-P strategy*, each *P* served by one and up to seven principles globally known as *the 14 principles of the Toyota Way*:

1° P - Philosophy:

1. Base your management decisions on a long-term philosophy, even at the expense of short-term financial goals.

2° P - Process:

2. Create a continuous process flow to bring problems to the surface.
3. Use *pull* systems to avoid overproduction.
4. Level out the workload, and work like the tortoise, not the hare (*heijunka*, 平準化).
5. Build a culture of stopping to fix problems, to get quality right the first time (*jidoka*, 自働化).
6. Standardised tasks and processes are the foundation of continuous improvement and employee empowerment.
7. Use visual controls so no problems are hidden.
8. Use only reliable, thoroughly tested technology that serves your people and process.

3° P - People and Partner:

9. Grow leaders who thoroughly understand the work, live the philosophy and teach it to others
10. Develop exceptional people and teams who follow your company's philosophy.
11. Respect your extended network of partners and suppliers by challenging them and helping them improve.

4° P - Problem solving:

12. Go and see for yourself to thoroughly understand the situation (*genchi genbutsu*, 現地現物).
13. Make decisions slowly by consensus, thoroughly considering all options; implement decisions rapidly.
14. Become a learning organisation through relentless reflection and continuous improvement.

As it can be easily seen, most principles are common to TQM philosophy, starting from the attention and care to the customer and their requirements. Lean management stands apart from traditional quality application, because of its rigour and determination to achieve the goals of *zero* waste and defects, as well as the use of specific tools developed and perfected by Taiichi Ohno.

In order to make its reduction more effective, waste is split into three categories: *muda*, wastefulness; *mura*, irregularities of the workload; and *muri*, overload. Muda is further subdivided into seven types of waste: correction, motion, overproduction, conveyance, inventory, processing and waiting.

Womack Jones and Ross [15], the American scholars who translated the TPS into the lean production, define a simple strategy to be followed when starting to apply it to an organisation. First, define the value as perceived by the customer, and then identify how the value flows and is generated throughout production; optimise its pathway without interruptions and let the customer *pull*[15] the value. Finally, seek perfection.

Several tools have been developed to sustain the entire workflow and to realise the 14 principles, and they are usually represented in the *Toyota Production System house* (see Fig. 6.27), whose representation slightly varies between texts. It is worth mentioning few of these tools and principles, because their fame transcended the mere application of this methodology: value stream mapping (VSM), takt time, spaghetti chart, 5S and *one-piece flow*.

Value stream mapping is the tool to represent the process under analysis *as is*, from the beginning to the release of the product or service to the customer, and the way in which the value for the customer flows through it. It is then used to design the future state *to be*. The process activities are classified into three types: (1) value added, which the customer is willing to pay for, (2) non-value added, but necessary for the subsequent activities or for legal requirements, and (3) non-value added, which needs to be removed.

To complete the VSM, information must be added regarding how material and information flow through the process, when and how long the process operates on the product, what and how much inventory or buffer stock is needed along the production line, how customer and suppliers interact with the process, which transport is needed and what tools and IT support are used.

Figure 6.28 illustrates an example of VSM, drafted for a simple process aimed at extracting information by stimulating cells. This VSM is simplified as applied by T. Barnhart, who condensed his experience of lean in research in an interesting text [16]. Indeed, the application of VSM is still very rare in scientific research. Barnhart suggests to use three simple metrics for each process step: *cycle time*, i.e. the total amount of time needed to complete the operation; *process time*, i.e. the actual time an operator works on the process step; and *per cent complete and accurate %C/A*, indicating the percentage of a good outcome at the first run. The map illustrates how information and materials flow through the production line as well as some data about production, waiting and transport times and inventory, how resources are

---

[15] This is one of the most specific concepts of Lean: Customer requirements drive the production pace, which is optimised to respond as quickly as possible to customer requests, instead of producing on estimates and storing it in warehouse awaiting for customer's order.

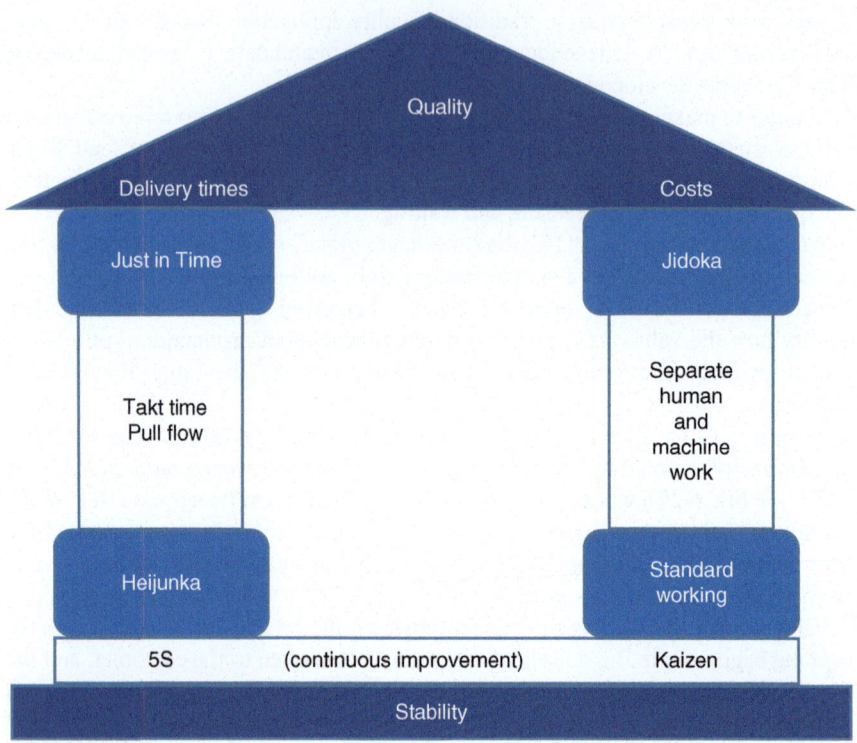

Fig. 6.27 The TPS house. At the top are the goals of the system: achieving the highest possible quality, with the lowest possible cost and within the shortest lead time. The goals stand on a two-pillar structure, just-in-time and Jidoka, and their related principles and tools. The basis is built by the continuous improvement principles as 5S and Kaizen, and the basic principle of stability of production

allocated and the opportunities for improvement. However, we shall not delve further into such details.

From the information gathered in the VSM, a key lean metric can be calculated: *takt time*. Takt time is the rate of production required to meet the customer's need, more specifically the available time divided by customer demand. Takt time is used to improve production levelling (*heijunka*), in order to avoid underuse of equipment

Fig. 6.28 Value stream mapping. A simplified version of the value stream mapping of a process for extracting information from stimulated cells is composed by (**a**) a scheme of information flow, (**b**) a scheme of materials/data flow and (**c**) the timeline showing waiting times and process times. The process starts upon a customer's request, which triggers the order to the cell line supplier and the planning of analysis and elaboration. Each process step is characterised by three main metrics: cycle time, process time and % C/A (% of complete and accurate outcome at the first run). More metrics may include information about number of operators, uptime (on-demand machine utilisation) or scrap rate. Amount of material (*inventory*) is indicated before and after each process step. *Process time* and *total lead time* are calculated at the end of the timeline

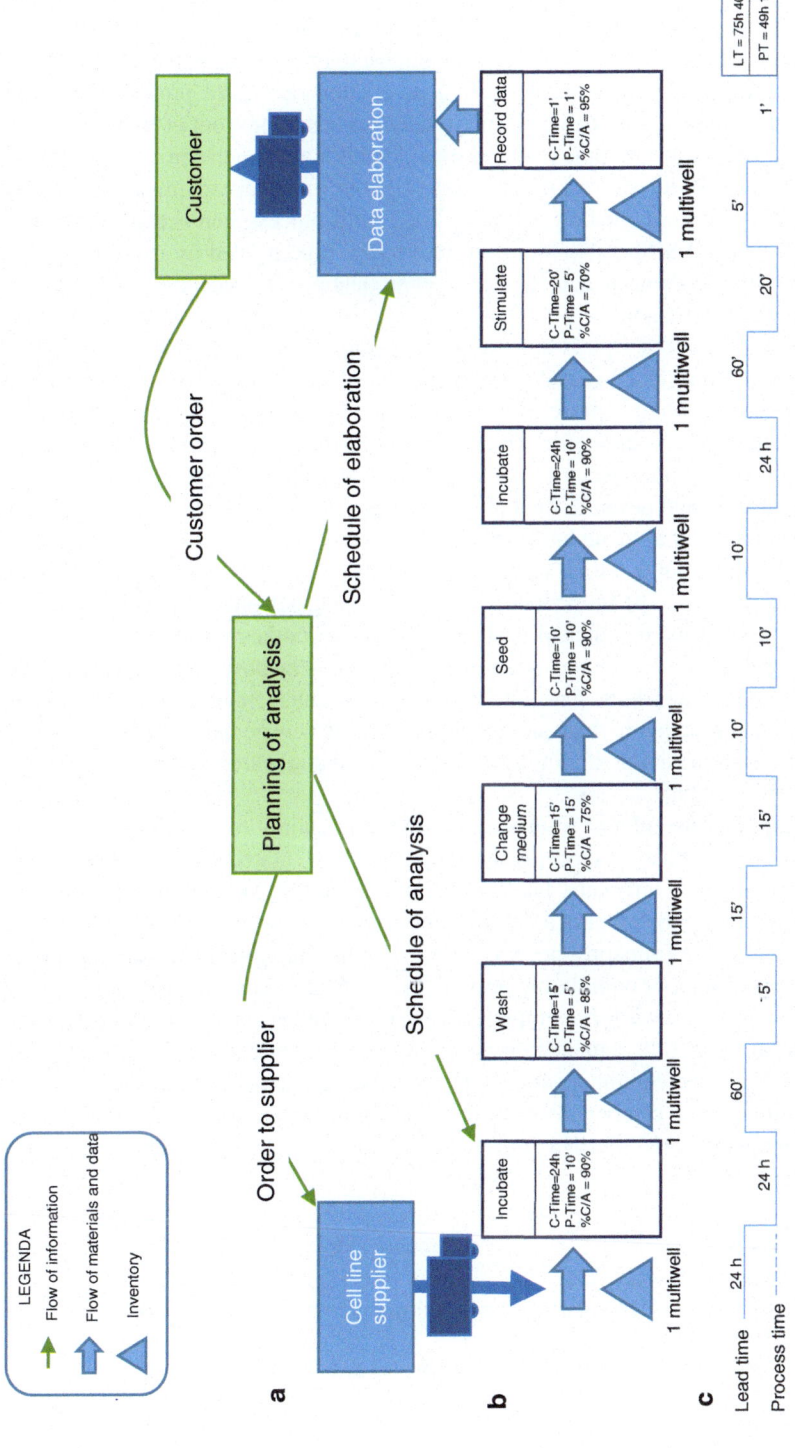

and the so-called bottlenecks[16] throughout the production line, and to understand where and why more resources should be added to the process to meet customer demand. Figure 6.29 illustrates the relationship between heijunka and takt time: as an example, the customer requires a production rate of 24 products/hour, corresponding to a takt time of 2.5 min. In the *as-is* state, three out of four process steps take longer than takt time (*before* series). In this situation, the process can only flow at the pace of the slowest process step (the fourth), i.e. approximately 13 products/hours (60 min divided by 4.4 min). The fourth process step is the bottleneck. An improvement action should aim to reduce the time needed by every process step below the takt time, to achieve the target production rate: the *after* series in Fig. 6.29 shows a hypothesis.

The *one-piece flow* is among the best known concepts of lean. According to Tahichi Ohno and his followers, the most efficient and effective production is achieved working a piece at a time and not, as usually done in manufacturing, processing them in batches. Figure 6.30 illustrates the comparison between the processing of a ten-piece batch and of a one-piece flow, showing a significant saving of time and costs. In his book *Creating a Lean &D System* [16], T. Barnhart, highlighting that in scientific research the outcome of production is new knowledge, points out that long sets of experiments delay the release of information that could otherwise be immediately reused. He suggests planning small, fast experiments in order to quickly confirm or refute hypothesis and tune further tests. Another positive outcome of the *one-piece flow* is the positive psychological effect that immediate results have on the researchers, enhancing the good working environment, motivating them and finally facilitating new good outcomes.[17] Again, this will result in a reduction in the waste of time and resources dedicated to achieve a scientific result.

Another useful tool to intercept waste, due mainly to superfluous movements and incorrect use of space, is the *spaghetti chart*. Starting from the floor plan of the workplace, all paths taken by people, material, sample and information to proceed with the process are drawn and measured. Improvement can be made for instance by eliminating process steps, moving instruments closer, changing the order of tasks, or eliminating rework. Figure 6.31 illustrates an example of a spaghetti chart drawn for a simple lab protocol—before—and a first, possible solution to reduce the path for the researcher during its execution, obtained just by moving closer equipment and benches—after. As an example, if a centrifuge used 20 times a day can be moved 10 m closer to the bench, this results in a 52 km per year saving! [18]. Further improvement may be achieved by intervening on the protocol itself.

---

[16]A workplace with reduced production capacity, which slows down the production flow, is called *bottleneck*.

[17]T. Barnhart refers to experimental results reported by T. Amabile in [17].

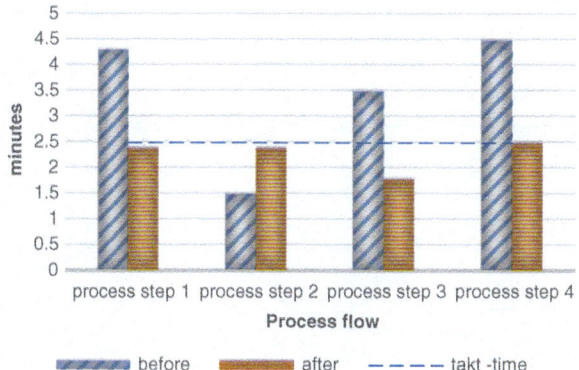

Fig. 6.29 Hejunka and takt time. The graph shows the processing time for four process steps for a production rate of 24 products/hour, before and after an improvement action. With a required production rate of 24 products/hour, corresponding to a takt time of 2.5 min (horizontal line), three out of four process steps take longer than takt time (*before* series). The process can only flow at the pace of the slowest process step (the fourth), i.e. approximately 13 products/hour (60 min divided by 4.4 min), with the fourth process step being the bottleneck. After an improvement action (*after* series), the time needed by every process step is reduced to below the takt time, thus achieving the target production rate

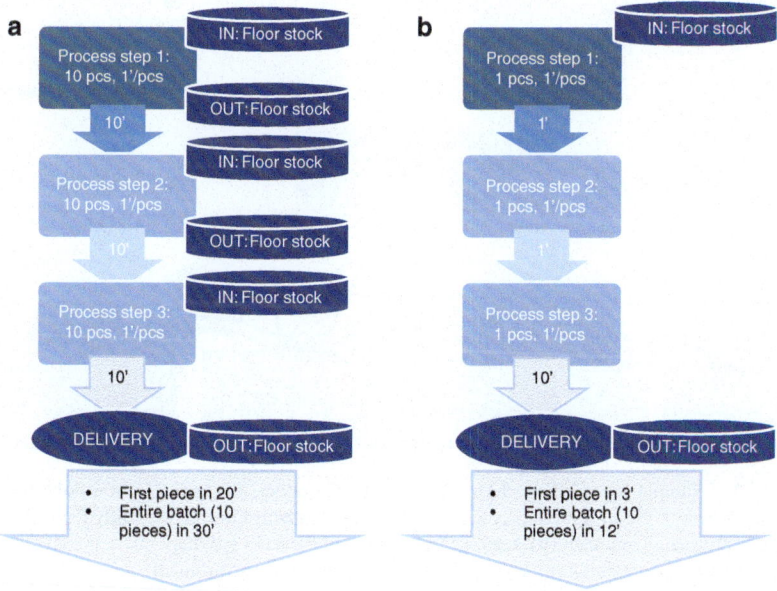

Fig. 6.30 Batch flow and one-piece flow. The process flow in (**a**) represents how a batch of ten pieces is managed in a three-step process. Each step needs floor stock in both input (*IN*) and output (*OUT*) and lasts 1 min for each piece; thus the entire batch needs a 10-min run to complete each step, because it is only delivered to the next step when all the ten pieces making up the batch are completed. The first piece, as part of a ten-piece batch, is ready after 10 min and an entire batch is completed after 30 min. The process flow in (**b**) represents the one-piece flow. Each single piece is worked and passed to the following process step in 1 min. The first piece can be delivered after 3 min and the entire ten-piece batch is completed in 12 min. There is no need for intermediate floor stock; thus the amount of material handled is dramatically reduced, significantly minimising costs and space required

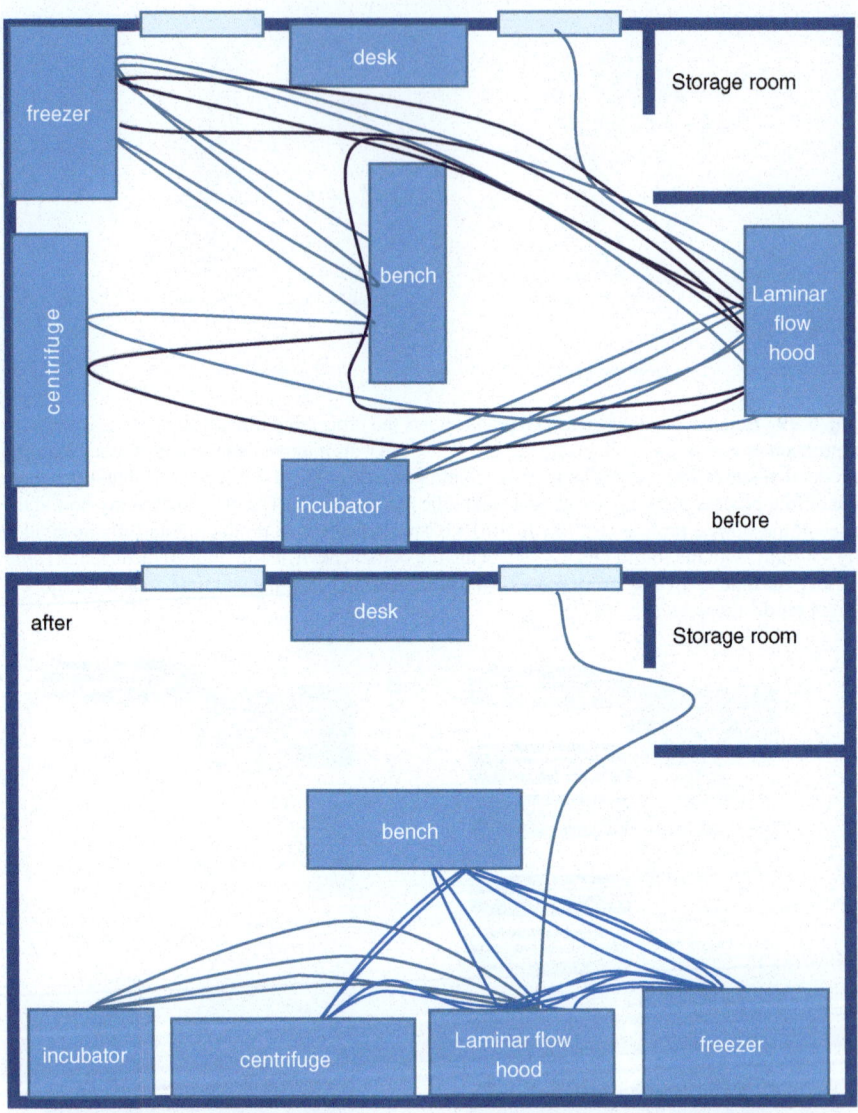

Fig. 6.31 Example of spaghetti chart for a simple laboratory protocol. (**a**) In a non-optimised laboratory layout, even a simple protocol forces the researcher to move frequently from one location or instrument to another. (**b**) After a careful study of the *as-is* spaghetti chart, the layout of the laboratory was changed to minimise the movements needed by the protocol

5S is a method for maintaining the workplace in safe, clean and optimised conditions so that the operators can execute their task, without any waste of time, material or space. The five phases must be executed in the order:

1. Seiri—Sort: Remove unnecessary objects, reduce waste.
2. Seiton—Set in order: Remove unnecessary objects, reduce waste and store rarely used items where they won't get in the way, but where they can be easily found.

3. Seiso—Shiny clean (also called *spik&span*): Determine and gain agreement on the desired level of cleanliness.
4. Seiketsu—Standardise: Make adhering to the new rules a constant habit for operators, and write and enforce standardised procedures.
5. Shitsuke—Sustain: Maintain the discipline and practice to improve.

For example, the 5S method helps achieving a simple but useful arrangement of storage: only the material needed for the current work is kept near the bench. Materials and tools of less frequent usage can be organised and labelled to facilitate their retrieval, and put in centralised storage to leave more room around the benches, while unused and obsolete materials can be periodically removed, freeing precious space and—last but not least—avoiding their improper usage.

Principles and tools of the lean management are already applied in analytics laboratories, where the production process is intended to provide analysis and elaboration of information instead of concrete products, but where the same problems encountered in manufacturing must be faced: productivity, lead time and profitability. As stated by Tom Reynolds, *When lean is applied properly in labs, productivity improvements of between 25 and 50% and/or lead time reductions of 80% are not unusual* [19]. A research laboratory can enjoy the same benefits, freeing resources to be dedicated to more challenging tasks and accelerating the scientific discovery process, in summary making the most from funding.

6.4.2   Six Sigma

The six sigma methodology was developed by Bill Smith, senior engineer and scientist, with the sponsorship of Bob Galvin, CEO of Motorola in the 1980s. It is based on concepts well known in the statistics: normal distribution by Carl Friedrich Gauss, three-sigma limit for a process by Walter Shewhart in the 1920s and several measurement standards developed during the decades that followed, as statistical process control (SPC) and CpK (a measure of how a process stays in control).

Six sigma is a set of techniques mainly based on statistical principles for improving the performance of processes, products or services, but it is also a philosophy and a goal: to be as perfect as practically possible. It can also be seen as an organisational model intended to improve effectiveness and efficiency with a project-based approach. It might seem a strange and polyhedral definition, somewhat heterogeneous, but in the following paragraphs we are going to see how these concepts can be reconciled as a single vision.

First, what is sigma? Identified with the Greek letter σ, a well-known concept for those who are familiar with statistics, is the standard deviation of a normally distributed population, i.e. the estimate of the variability of a population. For the sake of our discussion, it is sufficient to remember that in the area defined by $\pm 3\sigma$ around the mean stands 99.73% of the population of a Gaussian distribution (see Fig. 6.32). If the upper specification limit (USL) and the lower specification limit (LSL) of a process are set at the points of $\pm 3\sigma$, the process will have a defect rate of approximately 0.3%, which is unacceptable for most processes (see Box 6.3).

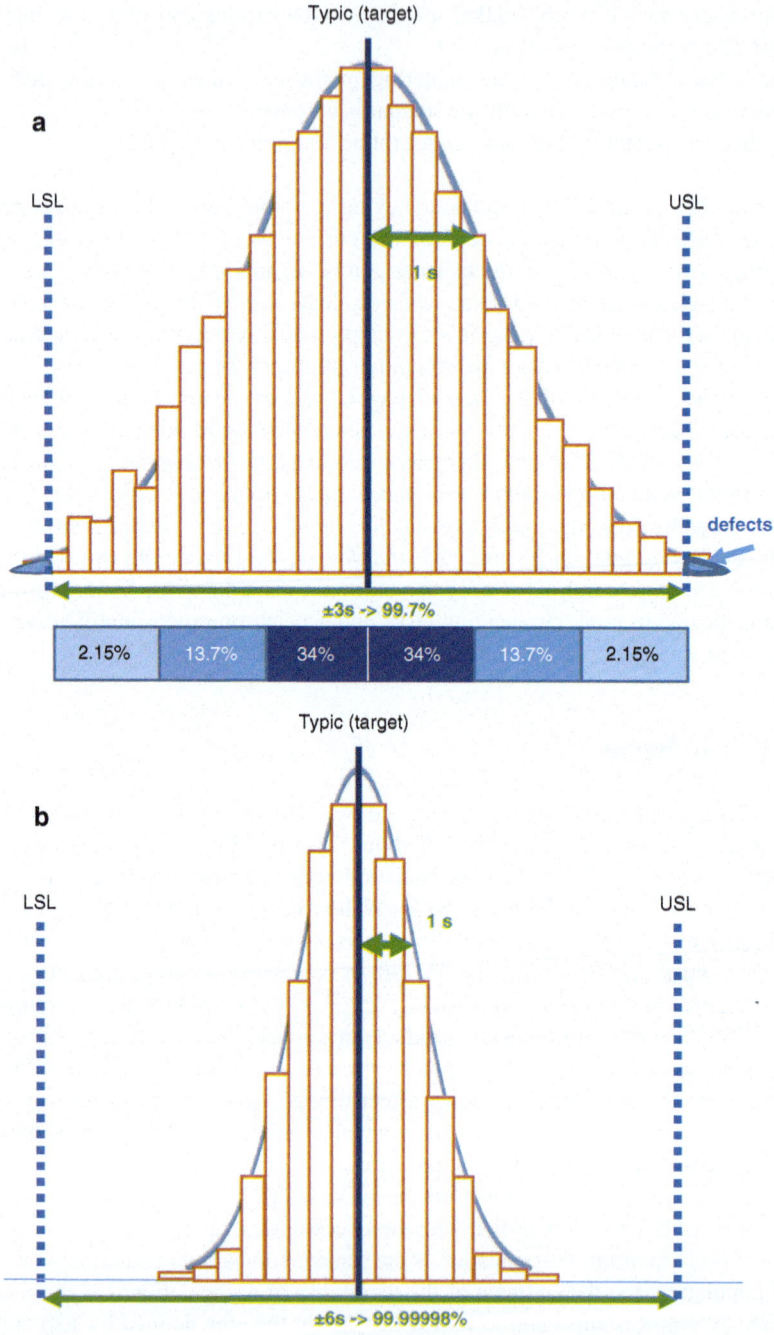

Fig. 6.32 Normal distribution and σ. Process data are normally distributed around the *typical (target)* value. The process limits USL (Upper Spec. Limit) and LSL (Lower Spec. Limit) are indicated with *dotted vertical lines*. The *dashed line* indicates the standard deviation (σ) of the process. The 3-σ breadth of the data distribution in (**a**) is within the process limits, but the greyed queues stretch beyond the process limits and generate *defects*. The distribution in (**b**) refers to a 6-σ process: no appreciable amount of data is out of bounds

The six sigma methodology uses the defects per million opportunities (DPMO), to measure the defect rate, and a correction factor to take into account the limitation in inaccuracy of the model (*σ-shift*). Bearing this in mind, Table 6.10 illustrates the correspondence between defect rate and number of sigma included in the specification range (USL to LSL). As it can be easily seen, a 3σ corrected with the σ-shift gives a defect rate of 66811 DPMO (6.68%), while a 6σ process can run at 3.4 DMPO (0.00034%). This means that a six sigma process generates 3.4 defects every million units produced.

Box 6.3 Why 3 or 4σ Are Not Sufficient?
4σ is equivalent to a yield of 99.9937% that means:

- 4000 wrong medical prescriptions every year
- More than 3000 infants accidentally falling from the arms of nurses or doctors in a year
- Every day 2 too long or too short landings at Chicago airport
- Every day our drinking water would be unsafe for about 15 min
- Every week there would be 5000 surgical operations failed to some extent
- Every month electricity would not be available for almost 7 h

---

Table 6.10 also shows another positive outcome of the six sigma method: approaching higher levels of sigma, an impressive economic result can be achieved regarding the costs of non-quality, by means of reduction of scraps, reworking and other wastes related to a high defect rate.

However, the sigma level scale is not linear. Improving from 2 to 3σ means a five times order of magnitude reduction in defects, but improving from 3 to 4σ is a tenfold reduction in the number of defects. This is how the methodology translates quantitatively the generic goal *to be as perfect as practically possible*.

Table 6.10 Defect rate and sigma level

Sigma level	% Yield (theoretical)	DPMO	% Yield (σ-shift corrected)	DPMO (σ-shift corrected)	Cost of non-quality (% on sales)
±2	95.45	45 500	69.15	308 537	30–40
±3	99.73	2 700	93.32	68 807	20–30
±4	99.9937	63	99.38	6 210	15–25
±5	99.999943	0.57	99.977	233	5–15
±6	99.999998	0.002	99.9997	3.4	Less than 1

For each sigma level (first column), the theoretical yield and the related *defects per million opportunities* (DMPO) rates are shown (second and third columns). The following two columns indicate the same parameters corrected by the *σ-shift* and the last one presents the costs of non-quality as a percentage on sales

To achieve the result of reducing the variability of each process to the 6σ level, a corrective approach is not enough: the methodology suggests focussing on the customer requirements to identify the performance characteristics of a process or a product that are critically important for the customer. In other words, paying attention to the voice of the customer (VOC) allows determining the *critical to quality* (CTQ). CTQs, which must be measurable, are the characteristics targeted by the improvement process and which must be addressed, with a problem-solving approach called DMAIC: this acronym stands for:

- Define
- Measure
- Analyse
- Improve
- Control

There are two more ingredients to be considered: people and organisation. More specifically, an effective improvement plan needs the commitment of upper management and it can be consolidated by means of an organisational structure with different levels of involvement and responsibility. That is why six sigma is also a company organisational model. Improvement is based on different formalised projects. For each project a *project champion* is appointed; usually this is an executive or a senior manager with enough authority to ensure resources, time and priority to the project. He or she selects the *project leader*, who is responsible for planning, leading and executing the project, and the members of the team, in agreement with the project champion. The project champion is also in charge of approving the project charter and any possible improvements proposed by the team. As in a matrix-type organisation, there is an expertise hierarchy other than that of project management, which starts from the lowest level consisting of the *white belts*, who own basic concepts about quality and process thinking, and the second level of the *yellow belts*, who undergo a formal training on DMAIC, VOC analysis, process mapping and root cause analysis. Further levels comprise the *green belts*, experts in data analysis and DMAIC projects; the *black belts*, who are full-time six sigma specialists, able to lead complex projects and to master methodologies as inferential statistic and DoE; and on the highest level the *master black belts*, who, thanks to their experience and deep knowledge of the methodology and related tools, serve as six sigma trainers and coaches to the rest of the organisation.

6.4.3   Applicability of Lean Management and Six Sigma in Basic Biomedical Research

What about the results that companies applying six sigma and lean management declare to have achieved. One of the most popular websites spreading the culture of six sigma [20] reports a thorough survey on actual data provided by the major companies following a six sigma programme, indicating that significant savings, varying from 1.2 to 4.5%, can be achieved. The same result cannot be literally translated to scientific research. Scientific research differs from manufacturing for the intrinsic difference between the products and the processes employed, not to mention the difference between the productive and creative environments and the characteristic of *high-variability and low-volume environment* inherent to research. Carleysmith, Dufton and Altria [21] investigated whether the lean and the six sigma methodologies could also benefit pharmaceutical research and development. Their conclusions were that those methods *can be used to improve knowledge management and teamwork, and to improve all routine aspects of the overall operation. The result is more time for scientists to innovate, and reduction of cycle times, which increases the speed of development. (…) Perhaps lean sigma does not promote creativity per se, but it certainly supports efficient problem definition, problem solving and dissemination of ideas.*

Barnhart [16] reports sixfold improvement in the speed of scientific discovery for a research team, freeing up to 50% of the time previously devoted to management.

According to Schweikhart and Dembe, six sigma techniques have been recommended as a means to avoid cross contamination of cell lines [14]. Hollensead et al. outline potential uses of lean, six sigma and other quality assurance practices to reduce laboratory errors in a host of disciplines including molecular biology, cytology, microbiology and pathology [22]. Lean and six sigma have also been directed towards quality assurance in pharmaceutical laboratories and production facilities, as reported among others by Carleysmith et al. [21]. Moreover, they propose a Decalogue to transpose the lean principles to a R&D environment. These are outlined here below:

1. Reduce batch size: take smaller steps with faster feedback of information.
2. Reduce the waste of unwanted variability to increase the capacity for desired variability, producing valuable information.
3. Focus on flow: Monitor flow of work and build-up of queues, and respond adaptively rather than preplanning everything far in advance. This approach tends to average out the arrival of technical problems and use of resources.
4. Pull, do not push. Rather than strategy driving roadmaps, driving plans driving monthly resourcing, use project pull to allocate resources frequently, giving shorter cycle times.
5. Ensure fast feedback of new information, so that scientists can control rapid development.
6. 'User requirements' are not as stable as in manufacturing: goals must adapt rapidly to all new information.
7. Invest in sufficient flexibility of people in R&D so that bottlenecks can be relieved.
8. Achieve adequate failure rates. An experiment generates knowledge most efficiently when its probability of success is 50%. Acceptable failure rates create less waste than trying to *do it right first time.*
9. Understand the economics of waste. In R&D, expenses are low compared with the cost of cycle times of months or years where people are the dominant cost. Focus on reducing cycle time even if this entails additional expense.

10. Control the right parameter—understand the critical path. For example, avoid local maximisation of the efficiency of support groups as this can reduce responsiveness and increase the overall cycle time.

More recently, various examples of lean applied to scientific research can be found, mainly in the United States and the UK. In Sandia, a US Government company engaged for national security in various fields including bioscience, lean programmes are run for more than 8 years, while several training courses and consultancies are already offered in the market to scientists who are willing to complement their scientific method with the principles of lean management and six sigma. A short research on the Web reveals a number of publications supporting the creation of a lean scientific environment—mainly in the pharmaceutical R&D. These may well become strong references for those who want to experiment with these methods in basic research.

In conclusion, we can retain the point of Schweikhart and Dembe [14] who, when discussing the complexity that biomedical research requires in terms of interrelations among disciplines and research entities, say: *It is naïve to believe that such a comprehensive view of biomedical research can be achieved without systematic tools and conceptual models for planning, understanding, analysing and implementing the diverse processes required for effective clinical and translational research. That is exactly the potential role that lean and six sigma are intended to serve. Those methodologies have been developed and honed in the equally complex environment of manufacturing and systems engineering, where quality, precision and customer uptake are as critical to overall project success as they are in biomedical research.*

References

1. A Guide to the Project Management Body of Knowledge (PMBOK® Guide). Project Management Institute (PMI). 2017.
2. ISO 21500:2012 guidance on project management.
3. IEC 60812:2006. Analysis techniques for system reliability—procedure for failure mode and effects analysis (FMEA).
4. AIAG. Potential failure mode and effects analysis (FMEA). 2008.
5. SAE J1739 potential failure mode and effects analysis in design (Design FMEA), Potential failure mode and effects analysis in manufacturing and assembly processes (Process FMEA). 2009.
6. Lanati A, et al. A design of experiment approach to optimize an image analysis protocol for drug screening. In: Zazzu V, Ferraro MB, Guarracino MR, editors. Mathematical models in biology. Heidelberg: Springer; 2015. p. 65–84.
7. Mancinelli S, et al. Applying design of experiments methodology to PEI toxicity assay on neural progenitor cells. In: Zazzu V, Ferraro MB, Guarracino MR, editors. Mathematical models in biology. Heidelberg: Springer; 2015. p. 45–63.
8. Antony J. Design of experiments for engineers and scientists. Amsterdam: Elsevier Science & Technology Books; 2003.
9. Montgomery D. Design and analysis of experiments. 8th ed. New York: Wiley; 2012.

10. Wang H. [Bio-Glossary]. Biological replicates vs technical replicates. 2012. https://hong-langwang.wordpress.com/2012/04/24/%E3%80%90bio-glossary%E3%80%91biological-replicates-vs-technical-replicates/. Accessed 15 Sep 2017.
11. Anderson MJ, Whitcomb PJ. DOE simplified: practical tools for effective experimentation. 3rd ed. Boca Raton: CRC Press; 2015.
12. FDA. Guidance for industry. 2012. https://www.fda.gov/drugs/guidancecomplianceregulatoryinformation/guidances/ucm313087.htm. Accessed 31 Oct 2017.
13. Washington State: lean in Washington. 2013. http://www.results.wa.gov/what-we-do/lean-washington. Accessed 15 Sep 2017.
14. Schweikhart SA, Dembe AE. The applicability of lean and six sigma techniques to clinical and translational research. J Investing Med. 2009;57(7):748–55. https://doi.org/10.2310/JIM.0b013e3181b91b3a.
15. Womack JP, Jones DT, Roos D. The machine that changed the world. New York: Free Press; 1990.
16. Barnhart T. Creating a lean R&D system. Boca Raton: CRC Press; 2013.
17. Amabile T. The progress principle: using small wins to ignite joy, engagement and creativity at work. Brighton: Harvard Business Review Press; 2011.
18. Siemens Healthcare: pursuing perfection through continuous improvement—an introduction to lean healthcare principles. 2015. https://static.healthcare.siemens.com/siemens_hwem-hwem_ssxa_websites-context-root/wcm/idc/groups/public/@global/@lab/@corelab/documents/download/mda1/ntgz/~edisp/a91dx-cai-151464-gc1-4a00_final-02556006.pdf. Accessed 15 Sep 2017.
19. Reynolds T. Leaning toward lean? 2010. http://www.labmanager.com/lab-design-and-furnishings/2010/07/leaning-toward-lean-#.WbueR8hJbIU. Accessed 15 Sep 2017.
20. iSixSIgma: six sigma costs and savings—the financial benefits of implementing six sigma at your company can be significant. 2008. https://www.isixsigma.com/implementation/financial-analysis/six-sigma-costs-and-savings/. Accessed 15 Sep 2017.
21. Carleysmith SW, Dufton AM, Altria KD. Implementing lean sigma in pharmaceutical research and development: a review by practitioners. R&D Manag. 2009;39(1):95–106.
22. Hollensead SC, Lockwood WB, Elin RJ. Errors in pathology and laboratory medicine: consequences and prevention. J Surg Oncol. 2004;88:161–18.

Conclusions

In a very interesting paper [1] regarding the risks that scientific research faces in the last evolution in the world of science, the author L. Ségalat draws a comparison between the last global financial crisis and the scientific world, identifying alarming similarities that, if not actually causing a comparable crash, may seriously endanger it. Although the core of Ségalat discussion is slightly outside the scope of this book, in the view of quality some considerations are worth noting. The worst problem highlighted by Ségalat is the harsh competitiveness in publishing and the priority that scientists set by preferring to publish in prestigious scientific journals rather than giving a substantial contribution to the progress of knowledge. Visibility is imperative to foster careers and to obtain funds and this pressure favours the publication of conclusions that are not supported by robust statistical analysis, the avoidance of negative results, in some cases data manipulation and even fraud: in brief, this pressure has led to an increase in incorrect or irreproducible results that make no contribution to science. Even if although the real solution to releasing the pressure on scientists is more complex and structured, involving policymakers and funding agencies, the author cites the quality approach as one of the tools that the scientific institutions and the scientific community should adopt to ensure adequate control on research procedures and consequently more robust and reliable outcomes.

Throughout the text, starting from the quality principles and basic concepts, we have made a long journey across international standards and references, methods and methodologies, best practice and examples. Our goal has been to collect skills, methods and information to support the researchers in their day-to-day job and to guarantee the reproducibility and the reliability of the results of their scientific research. We have also seen the importance of consistent data and analysis, well-structured procedures, organisation and planning, controls and verifications, and the importance of simple, validated and reliable methods to streamline analysis and elaborations. The human side has not been neglected, following the TQM principles that state *people are one of the most important assets of an organisation* and as such they should be given optimal conditions to allow them to contribute at their best to the common goals. The so-called soft skills, such as leadership and team work,

© Springer International Publishing AG, part of Springer Nature 2018
A. Lanati, *Quality Management in Scientific Research*,
https://doi.org/10.1007/978-3-319-76750-5

communication and time management, are most often overlooked in the training of scientists, a good opportunity lost, since soft skills are of paramount importance to create, maintain and allow the flourishing of a human environment that favours the generation of good outcomes, especially intellectual ones such as scientific results.

It is time for life science research to generate its own international references and standards. Non-clinical research and clinical research—together with drug manufacturing and pharmacovigilance—have their widespread and internationally accepted good practice. In contrast, even the simplest discussion in an international arena regarding good research practice in life sciences is still to come. As discussed in the text, a lot of knowledge is already available: tools and methodologies have often been borrowed from *adjacent* fields, such as aeronautics, chemical, semiconductors and automotive industries, where the challenges of reliability have long since been addressed because of economic pressure or safety needs. Why not then follow their lead and pursue the synergy between the different fields when the goal is so similar? Most of quality concepts and methods are exploitable, with slight adjustments, by the scientific method; furthermore, adopting these methodologies will favour the translation of results from academia to industry and will foster the interdisciplinary cooperation, which is always a source of synergy and innovation.

The biggest obstacle remains the attitude of some researchers, who are very confident in the scientific method (whose validity is not in dispute) and in the peer review (as seen, not so flawless), and do not perceive the relevance of quality approaches and tools to guarantee the reliability of data—accurate, reproducible and traceable—thus effectively and honestly contributing to scientific progress (see Box A.1).

Box A.1: Goodness of Data
Experiments really are getting more complex, generating enormous quantities of data. But these data are only as good as the last calibration of the myriad pieces of equipment used to collect them, the quality controls on incoming materials and the rigorous tracking and reporting of both successful and failed experiments to allow for root cause analysis.
    Bill Frezza [2]

A voluntary commitment to a quality-based code of conduct in research is always possible, and some universities and councils have already adhered to it. Research institutions and single laboratories, wanting sound science and robust results, may also consider adhering to the quality code. References for the drafting of codes for quality research can be found in the WHO-TDR Handbook on QPBR [3], RQA Guidelines [4] and other university GRPs.

But a code of conduct however, is not enough to change mindsets. Following the GRPs may seem to scientists to be a time-consuming activity, alimenting fears of regulatory bureaucracy and stealing valuable resources from experimental work. Those who have experienced the quality approach from the beginning of their scientific career knows that it is a very fruitful investment for the future, paying off not only with robust results and substantial contributions to science, but also saving time and resources which would be otherwise spent in rectifying and reworking. That is why it is fundamental to develop young scientists in the culture of quality and teach them from very early on to work following GRPs.

Finally, policymakers and granting agencies have an essential role: Quality approaches should be not only suggested, but also made mandatory for grant-funded projects. A groundbreaking example is the JCoPR, which must be adopted by research institutions in the UK to have access to public funding.

My goal in writing this book, which has taken several years and is a distillation of my most important experiences in the field of quality as applied to scientific research in the life sciences, was to provide a reference to researchers who believe that good science cannot survive without both a strong methodological approach and enlightened management.

References

1. Ségalat L. System crash. EMBO Rep. 2010;11(2):86–9. https://doi.org/10.1038/embor.2009.278.
2. Frezza B. What's fueling our growing loss of faith in big science? 2013. http://www.bio-itworld.com/2013/1/11/whats-fueling-our-growing-loss-faith-big-science.html. Accessed 15 Sept 2017.
3. WHO: WHO-TDR handbook: quality practices in basic biomedical research. 2010. http://www.who.int/tdr/publications/documents/quality_practices.pdf?ua=1. Accessed 13 Sept 2017.
4. RQA working party on quality in non-regulated research. Guidelines for quality in non-regulated scientific research booklet. RQA. 2008.

The manufacturer's authorised representative in the EU is Springer
Nature Customer Service Centre GmbH, Europaplatz 3, 69115 Heidelberg,
Germany. If you have any concerns regarding our products, please
contact ProductSafety@springernature.com

Printed and bound by CPI Group (UK) Ltd, Croydon, CR0 4YY
27/04/2026
02097605-0001